10 Minnesota MCA Grade 7 Math Practice Tests

The Ultimate Test Prep Collection with Answer Explanations

Dr. A. Nazari

10 Practice Tests

Grade 7 Math — Engineered for Mastery

Welcome to **The Architect's Workshop.**

Ten practice tests. Ten floors of a tower you're about to build. Each test lays another level of understanding — from foundation to capstone.

- **Foundation (Tests 1–3):** Learn the blueprints
- **Framework (Tests 4–6):** Build structural strength
- **Finishing (Tests 7–9):** Refine under pressure
- **Capstone (Test 10):** The final inspection

Precision. Practice. Perfection.

> **"** Every great structure
> is built one floor at a time.
> Ten levels of practice makes
> your math rock-solid. **"**

✐ The Architectural Plan ✐

A 4-phase construction plan for complete mastery

Phase I: Foundation (Tests 1–3)

Untimed. Explore the test format and question types. Read every answer explanation after each test. Goal: understand the blueprint before you build.

Phase II: Framework (Tests 4–6)

Add a timer (70 minutes). Focus on the topics that gave you trouble in Phase I. Practice showing complete work and labeling units. Goal: build structural strength.

Phase III: Finishing (Tests 7–9)

Full timed conditions (60 minutes). Simulate the real exam. Review only the questions you missed — don't re-study what you already know. Goal: refine under real pressure.

Phase IV: Capstone (Test 10)

Your final inspection. Full exam conditions. This is the capstone — compare with Test 1 and measure your entire growth arc. Goal: prove your mastery.

📋 Blueprint Specifications

✐ **Multiple Choice** — select the single best answer

✐ **Short Answer** — show your process

✐ **Multi-Select** — choose ALL that apply

✐ **Open Response** — explain and justify

📐 Engineering Principles 📐

Precision techniques for building every solution

📐 The D.R.A.F.T. Method

D **Define** What exactly is the question asking? Write it in your own words.

R **Retrieve** Pull out given information. List numbers, units, and relationships.

A **Assemble** Choose the right formula or strategy. Set up the equation or proportion.

F **Figure** Compute step by step. Show every operation. Label units.

T **Test** Does the answer make sense? Re-read the question. Verify the units.

📐 Seven Precision Rules

1. **Read twice, solve once.** The first read gives context; the second reveals what to compute. 2

3. **Estimate before calculating.** A quick mental approximation catches major errors. 4

5. **Track your signs.** Rational number operations are the #1 error source in Grade 7. 6

7. **Never submit blanks.** On open response, even a partial setup can earn credit.

🕐 Timing Blueprint

Tests 1–3: **Untimed** (learn the blueprints) ＞ Tests 4–6: **70 min** (build speed) ＞ Tests 7–10: **60 min** (exam conditions)

Every great structure starts with a solid plan. Study the blueprints, learn from each test, and build something extraordinary.

Find more at
ViewMath.com/MN-Grade7

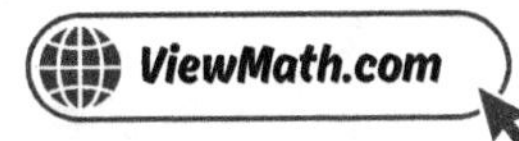

🧰 The Drafting Toolkit 🧰

Prepare your station before every construction session

⚙ Required Equipment

✏	**Precision Pencils**	Two or more #2 pencils, kept sharp. A dull tool produces dull work.
🧽	**Quality Eraser**	Clean, smudge-free corrections. Architects revise — so will you.
🗒	**Scratch Paper**	Your drafting workspace. All calculations happen here first.
📏	**Ruler / Protractor**	Precision tools for scale drawings, angle measures, and geometry.
⏱	**Timer**	Begin using from Phase II (Test 4) onward.
🔈	**Quiet Workspace**	A clean, well-lit station free from distractions.

✅ Approved Materials

- ✔ Pencil and eraser
- ✔ Scratch paper (provided on test day)
- ✔ Ruler or protractor (if specified)
- ✔ Reference sheet (if provided)

🚫 Restricted Materials

- ✖ Calculator (unless specified)
- ✖ Electronic devices
- ✖ Notes or reference materials
- ✖ Communication with others

❤ Project Manager's Guide (for Parents & Guardians)

Ten practice tests provide the most comprehensive preparation available. The 4-phase structure (Foundation → Framework → Finishing → Capstone) ensures skills build progressively and durably.

Keys to a successful build:

- Space tests 2–3 days apart — never more than one per day
- Phases I–II: review answers together and discuss strategies
- Phases III–IV: let them work independently, then debrief results
- Compare Test 1 with Test 10 to celebrate the full construction arc

Find more at
ViewMath.com/MN-Grade7

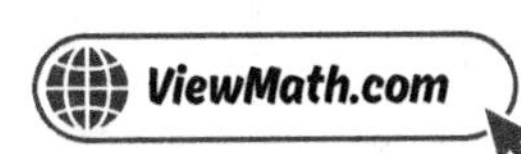

Math Reference Sheet

Symbol	Name	What It Means	
$(\ )$	Parentheses	Do this part first.	$(3+4) \times 2 = 14$
10^3	Exponent	Multiply the base by itself that many times. $10^3 = 1{,}000$	
$\dfrac{a}{b}$	Fraction	a parts out of b equal parts; also means $a \div b$.	
$\dfrac{7}{3}$	Improper Fraction	Numerator $\geq$ denominator.	$\frac{7}{3} = 2\frac{1}{3}$
0.45	Decimal	A number with a decimal point.	$0.45 = \frac{45}{100}$
$> \ < \ =$	Comparison	Greater than, less than, equal to.	$0.5 > 0.35$
$(3,5)$	Ordered Pair	A point on the coordinate plane: (x, y).	

Key Formulas

- **Volume of a rectangular prism:**

  $V = l \times w \times h$

- **Order of operations:**

  Parentheses → Exponents → Multiply/Divide → Add/Subtract

- **Powers of 10:**

  $10^1 = 10 \quad 10^2 = 100$

  $10^3 = 1{,}000 \quad 10^4 = 10{,}000$

- **Fraction as division:**

  $\frac{a}{b} = a \div b$

⊞ Place Value Chart

Millions	1,000,000	**Decimals**	
Hundred-Thousands	100,000	Tenths	0.1
Ten-Thousands	10,000	Hundredths	0.01
Thousands	1,000	Thousandths	0.001
Hundreds	100		
Tens	10		
Ones	1		

Each place is **10**× the place to its right, and $\frac{1}{10}$ of the place to its left.

📖 Key Math Vocabulary

- **Sum** — the result of addition
- **Difference** — the result of subtraction
- **Product** — the result of multiplication
- **Quotient** — the result of division
- **Remainder** — what's left over after dividing
- **Factor** — a number you multiply
- **Expression** — numbers and operations without =
- **Equation** — a math sentence with =
- **Numerator** — the top number of a fraction
- **Denominator** — the bottom number of a fraction
- **Mixed number** — a whole number + a fraction
- **Equivalent fractions** — fractions with equal value
- **Decimal** — a number written with a decimal point
- **Volume** — the space inside a 3D shape
- **Coordinate plane** — a grid with x and y axes
- **Ordered pair** — (x, y) location on the plane

🔍 Word Problem Clue Words

- **Add** (+): total, altogether, combined, sum, increase, more than
- **Subtract** (−): difference, how many more, fewer, remain, decrease, left
- **Multiply** (×): each, every, groups of, times, product, per, of (with fractions)
- **Divide** (÷): share equally, split, each group, how many groups, quotient, per

Find more at
ViewMath.com/MN-Grade7

▦ *Multiplication Table* ▦

×	1	2	3	4	5	6	7	8	9	10	11
1	1	2	3	4	5	6	7	8	9	10	11
2	2	4	6	8	10	12	14	16	18	20	22
3	3	6	9	12	15	18	21	24	27	30	33
4	4	8	12	16	20	24	28	32	36	40	44
5	5	10	15	20	25	30	35	40	45	50	55
6	6	12	18	24	30	36	42	48	54	60	66
7	7	14	21	28	35	42	49	56	63	70	77
8	8	16	24	32	40	48	56	64	72	80	88
9	9	18	27	36	45	54	63	72	81	90	99
10	10	20	30	40	50	60	70	80	90	100	110
11	11	22	33	44	55	66	77	88	99	110	121

💡 How to Use This Table

To find **4 × 7**:

1. Find **4** in the left column (blue).
2. Find **7** in the top row (blue).
3. Follow the row and column until they meet: the answer is **28**!

Tip: You can also use this table for division! If you know $28 \div 4 = $?, find 28 in the 4's row. The column header gives you the answer: **7**!

🏢 Construction Log 🏢

Record each floor and watch your tower rise

Architect: _________________________________   **Project Start:** _________________

Phase I: Foundation — Floors 1–3

📐 **Floor 1**   Date: _________   Score: _____ / _____   %: _____   ⭐⭐⭐⭐⭐

📐 **Floor 2**   Date: _________   Score: _____ / _____   %: _____   ⭐⭐⭐⭐☆

📐 **Floor 3**   Date: _________   Score: _____ / _____   %: _____   ⭐⭐⭐⭐☆

Phase II: Framework — Floors 4–6

📐 **Floor 4**   Date: _________   Score: _____ / _____   %: _____   ⭐⭐⭐⭐☆

📐 **Floor 5**   Date: _________   Score: _____ / _____   %: _____   ⭐⭐⭐⭐☆

📐 **Floor 6**   Date: _________   Score: _____ / _____   %: _____   ⭐⭐⭐⭐☆

Phase III: Finishing — Floors 7–9

📐 **Floor 7**   Date: _________   Score: _____ / _____   %: _____   ⭐⭐⭐⭐☆

📐 **Floor 8**   Date: _________   Score: _____ / _____   %: _____   ⭐⭐⭐⭐☆

📐 **Floor 9**   Date: _________   Score: _____ / _____   %: _____   ⭐⭐⭐⭐☆

Phase IV: Capstone — Floor 10

📐 **Floor 10**   Date: _________   Score: _____ / _____   %: _____   ⭐⭐⭐⭐⭐

Certification Level: Apprentice (0–39%) Technician (40–59%) Engineer (60–79%) **Master Architect (80–100%)**

Record your certification level after each test!

🔧 Under Construction

- ____
- ____
- ____
- ____

✓ Construction Complete

- ★ ____
- ★ ____
- ★ ____
- ★ ____

Find more at
ViewMath.com/MN-Grade7

⭐ Table of Contents ⭐

Here's what we'll explore together!

 Let's learn and have fun!

Practice Test 1

30 Questions

✏️ Before You Start ✏️

- ✓ **Read each question carefully** before choosing your answer.
- ✓ **Show your work** on scratch paper when you need to.
- ✓ **Skip hard questions** and come back to them later.
- ✓ **Check your answers** when you're done.
- ✓ **Take your time** — there's no rush!

⭐ You've Got This! ⭐

Do your best and show what you know!

1. The table shows a proportional relationship. Find k and the missing value.

x	y
3	7.5
?	15
10	25

Your Answer:

2. Two proportional lines are graphed below. Which statement is correct?

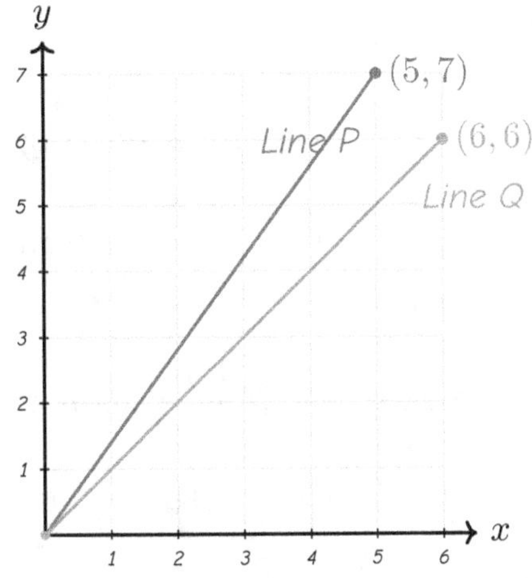

(A) Line Q has the greater unit rate because it is longer

(B) Line P has the greater unit rate because it is steeper

(C) Both lines have the same unit rate

(D) Line Q has the greater unit rate because $6 > 5$

Find more at
ViewMath.com/MN-Grade7

ViewMath.com

3. The graph below compares the rates of two workers packaging boxes. Which worker is faster, and by how many boxes per hour?

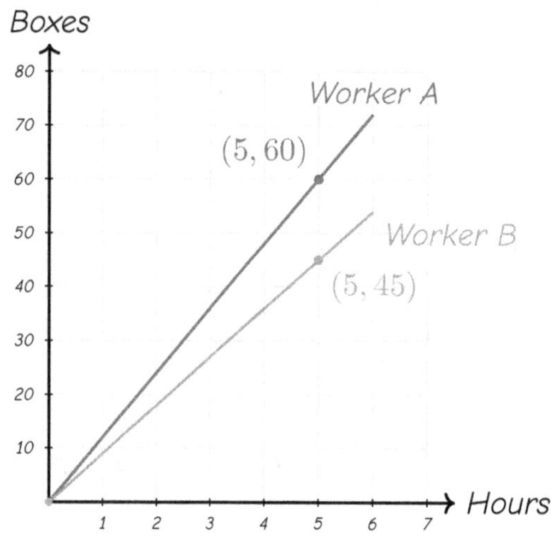

  (A) Worker A, by 3 boxes/hr   (B) Worker B, by 3 boxes/hr

  (C) Worker A, by 15 boxes/hr   (D) They work at the same rate

4. A store has 300 items in stock. If 60% are on sale, how many items are on sale?

  (A) 120   (B) 150

  (C) 180   (D) 200

5. The diagram shows how a salesperson's earnings are calculated. What are the total earnings?

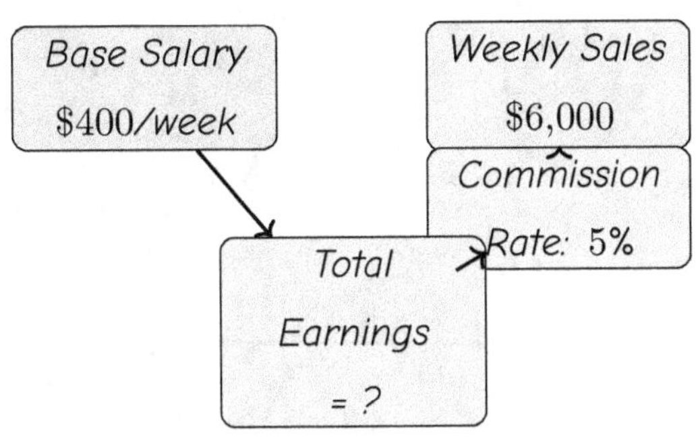

(A) $430

(B) $700

(C) $6,400

(D) $6,700

6. A jogger estimated running 5 miles but actually ran 4.8 miles. What is the percent error (rounded to the nearest tenth)?

Your Answer:

7. The table below shows the monthly fees at four financial institutions. Which institution has the lowest annual fee?

Institution	Monthly Fee
Big Bank	$8
Town Credit Union	$2
City Bank	$5
Online Bank	$0

(A) Big Bank

(B) Town Credit Union

(C) City Bank

(D) Online Bank

8. A football team gains 8 yards on one play and loses 8 yards on the next play. What is the net result?

   (A) 16 yards gained

   (B) 8 yards gained

   (C) 8 yards lost

   (D) 0 yards

9. Evaluate $\frac{x^2-1}{4}$ when $x = 5$.

   Your Answer:

10. Simplify $\frac{1}{2}x + \frac{3}{4}x - 2$.

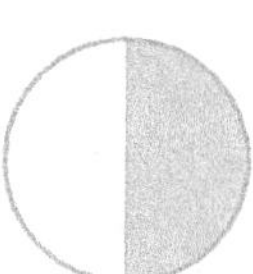

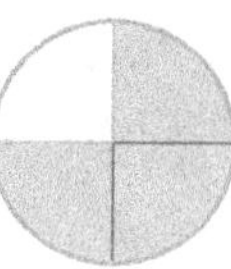

   Your Answer:

11. The diagram shows a triangle with a perimeter of 39 cm. What is the value of x?

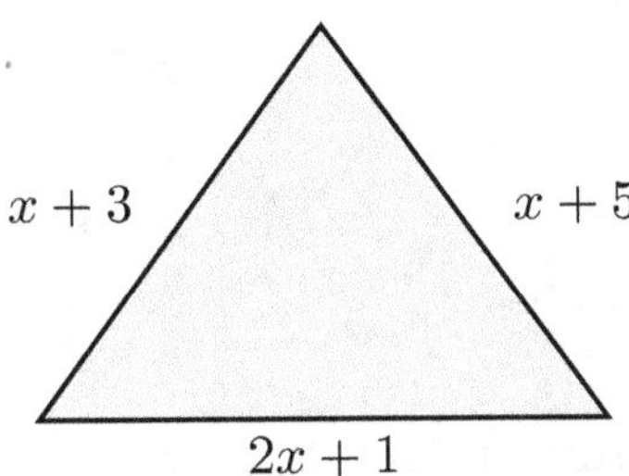

(A) $x = 7.5$

(B) $x = 10$

(C) $x = 6$

(D) $x = 8$

12. Four friends go to a museum. Each person pays for a ticket and a \$3 audio guide. The total for all four people is \$68. The equation $4(t + 3) = 68$ represents this situation, where t is the ticket price. Use the model below to find t.

Your Answer:

13. A store marks down a jacket by $\frac{1}{4}$ of its original price p, then subtracts an additional \$8. The sale price is \$37. Find the original price.

Your Answer:

14. *Solve* $2x + 3 > 11$.

(A) $x > 4$

(B) $x > 7$

(C) $x < 4$

(D) $x > 14$

15. What inequality is shown by a closed circle at -6 with shading to the right?

Your Answer:

16. You know two sides of a triangle are 7 cm and 10 cm, and the angle between them is 45°. How many different triangles can you draw?

(A) None

(B) Exactly one

(C) Exactly two

(D) Infinitely many

17. A triangle has sides 11, 13, and 20. Does it satisfy the triangle inequality?

Your Answer:

18. What shape is the cross-section when a cone is sliced vertically through its apex?

Your Answer:

19. Adjacent angles on a straight line must sum to:

(A) 90°

(B) 180°

(C) 270°

(D) 360°

Find more at
ViewMath.com/MN-Grade7

20. A circle has a diameter of 14 cm. What is its area? Use $\pi \approx \frac{22}{7}$.

(A) $44\ cm^2$

(B) $154\ cm^2$

(C) $308\ cm^2$

(D) $616\ cm^2$

21. A figure is made of a rectangle 12 cm by 5 cm with a right triangle removed. The triangle has legs 4 cm and 5 cm. What is the area of the remaining figure?

(A) $40\ cm^2$

(B) $50\ cm^2$

(C) $60\ cm^2$

(D) $70\ cm^2$

22. A triangular prism has a triangular base with base 6 cm and height 4 cm, and the prism is 10 cm long. The three rectangular faces have widths 6 cm, 5 cm, and 5 cm. What is the total surface area?

(A) $184\ cm^2$

(B) $160\ cm^2$

(C) $124\ cm^2$

(D) $120\ cm^2$

23. A rectangular fish tank is 50 cm long, 30 cm wide, and 40 cm tall. What is the volume?

(A) $120\ cm^3$

(B) $6{,}000\ cm^3$

(C) $60{,}000\ cm^3$

(D) $600\ cm^3$

24. In a school of 600 students, a random sample of 60 found that 15 students walk to school. Predict how many students in the entire school walk to school.

(A) 100

(B) 150

(C) 200

(D) 250

Find more at
ViewMath.com/MN-Grade7

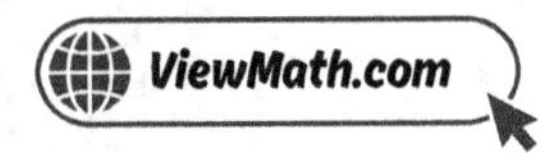

25. Which measure of variability is shown directly on a box plot?

(A) Mean absolute deviation (MAD)

(B) Standard deviation

(C) Interquartile range (IQR)

(D) Mean

26. The table compares two delivery companies' shipping times (in days).

	Company X	Company Y
Mean	3.5 days	3.2 days
Median	3 days	3 days
Range	4 days	8 days
IQR	1.5 days	4 days

A customer wants the most reliable delivery time. Which company should they choose?

(A) Company Y, because its mean is lower

(B) Company X, because its IQR and range are both smaller

(C) Company Y, because its range includes more possibilities

(D) They are equally reliable

27. A class of 40 students: 16 prefer dogs, 12 prefer cats, 8 prefer fish, 4 prefer birds. What percent prefer dogs?

(A) 16%

(B) 25%

(C) 40%

(D) 80%

28. A probability model has outcomes X, Y, and Z. $P(X) = 0.15$ and $P(Y) = 0.45$. What is $P(Z)$?

Your Answer:

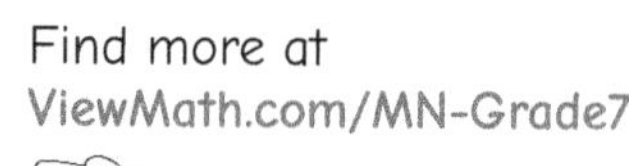

29. A spinner has 4 equal sections $(1, 2, 3, 4)$ and a die is rolled. What is the probability that both show a 3?

(A) $\frac{1}{10}$

(B) $\frac{1}{24}$

(C) $\frac{2}{10}$

(D) $\frac{1}{4}$

30. A spinner has a 60% chance of landing on blue. In a simulation of 80 spins, how many times would you expect blue to appear?

Your Answer:

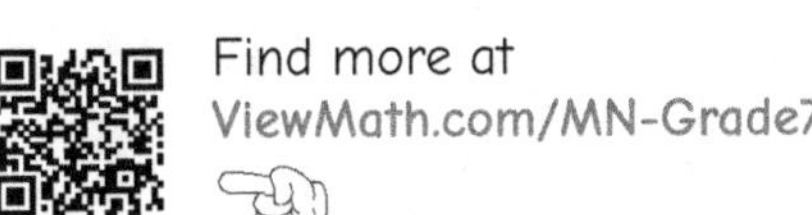

 # ★ *End of Practice Test 1* ★ 

Great job finishing the test!

📋 My Score

I got _____________ out of 30 questions right.

*Check your answers in the **Answer Key** at the back of the book.*

💡 *Review any questions you missed. That's how we learn!*

📊 Check Your Score Online!

Visit **ViewMath Academy** to enter your answers and see which topics you need to review. You can also explore lessons, take quizzes, track your scores, and save your progress!

viewmath.com/score/7.1.MN.16

Or go to viewmath.com/score and enter code: 7.1.MN.16

2

Practice Test 2

 30 Questions

---

✏️ Before You Start ✏️

✓ **Read each question carefully** before choosing your answer.

✓ **Show your work** on scratch paper when you need to.

✓ **Skip hard questions** and come back to them later.

✓ **Check your answers** when you're done.

✓ **Take your time** — there's no rush!

 You've Got This! 

Do your best and show what you know!

1. The table below shows a proportional relationship between gallons of gas and miles driven. What is k and what does it mean?

Gallons (x)	Miles (y)
2	54
5	135
8	216

(A) $k = 2$; uses 2 gallons per trip

(B) $k = 27$; drives 27 miles per gallon

(C) $k = 54$; drives 54 miles on 2 gallons

(D) $k = 135$; drives 135 miles on 5 gallons

2. The graph below shows the number of birdhouses a carpenter can build over time. What is the unit rate, and how many birdhouses can be built in 10 hours?

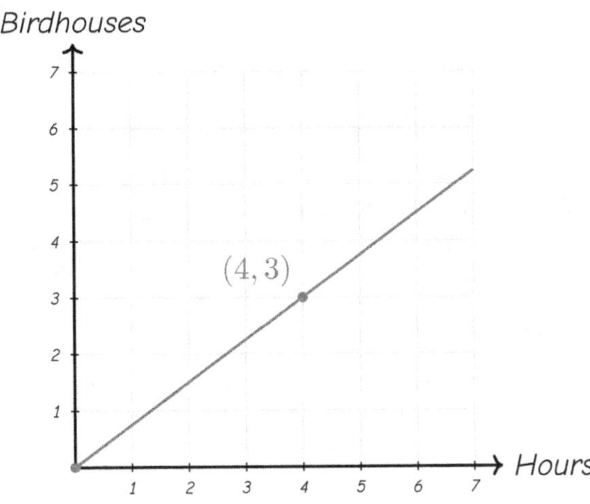

Your Answer:

3. The table shows how many batches of muffins a bakery can make and the total flour used. How many cups of flour are needed for 15 batches?

Batches	Flour (cups)
3	$7\frac{1}{2}$
6	15
9	$22\frac{1}{2}$
15	?

Your Answer:

4. A class has 32 students. If 75% passed the test, how many students passed?

(A) 20

(B) 24

(C) 25

(D) 28

5. A bill is $86 and you leave a $17.20 tip. What percent tip is that?

(A) 15%

(B) 17%

(C) 18%

(D) 20%

6. Two students estimated the number of beans in a jar. The actual count was 200. Student A guessed 180; Student B guessed 220. Who had the greater percent error?

(A) Student A

(B) Student B

(C) Same percent error

(D) Cannot be determined

Find more at
ViewMath.com/MN-Grade7

ViewMath.com

7. You charge \$300 on a credit card with an 18% annual interest rate. If you pay it off after 1 year, how much interest do you owe?

(A) \$18

(B) \$36

(C) \$54

(D) \$318

8. Find two integers whose absolute values are both 7.

Your Answer:

9. What is the value of $\frac{x}{2} - 4$ when $x = 10$?

(A) 3

(B) -1

(C) 1

(D) 9

10. Simplify $-2a + 5b + 7a - 3b$.

Your Answer:

11. Solve $8 = \frac{x}{3} + 2$.

(A) $x = 30$

(B) $x = 18$

(C) $x = 2$

(D) $x = 6$

12. Solve $2(x + 7) = -6$.

Your Answer:

Find more at
ViewMath.com/MN-Grade7

13. A number line shows a point at x. The equation $\frac{x}{4} + 1.5 = 5$ tells you how to find x. Solve and identify where x is on the number line.

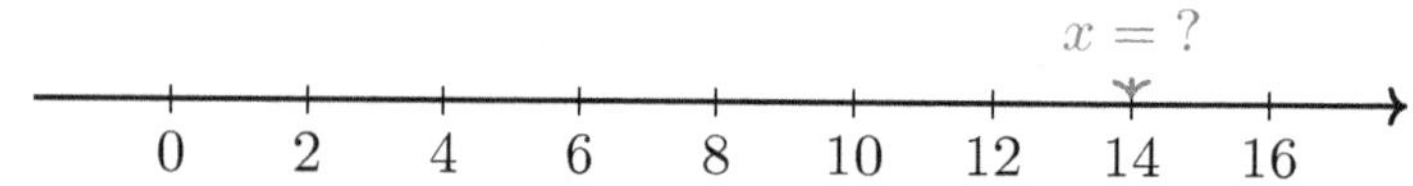

Your Answer:

14. Solve $\frac{x}{2} + 7 \leq 12$.

Your Answer:

15. Which inequality is shown by a closed circle at 7 with shading to the left?

(A) $x > 7$

(B) $x < 7$

(C) $x \geq 7$

(D) $x \leq 7$

16. You know all three angles of a triangle: 50°, 60°, and 70°. How many triangles can be drawn?

(A) None

(B) Exactly one

(C) Exactly three

(D) More than one (infinitely many)

17. Two angles of a triangle are 35° and 75°. What is the third angle?

(A) 60°

(B) 70°

(C) 80°

(D) 110°

18. A cube can be sliced to produce a hexagonal cross-section. How?

(A) By cutting parallel to the base

(B) By cutting diagonally through all six faces

(C) By cutting vertically through the center

(D) It is impossible to get a hexagon from a cube

19. Find the supplement of 43°.

Your Answer:

20. A circle has an area of 78.5 cm^2. Using $\pi \approx 3.14$, what is the radius?

(A) 4 cm

(B) 5 cm

(C) 10 cm

(D) 25 cm

21. A swimming pool is shaped like a rectangle 25 m by 10 m with a semicircle of diameter 10 m at one end. What is the total area? Use $\pi \approx 3.14$.

(A) 250 m^2

(B) 289.25 m^2

(C) 328.5 m^2

(D) 367.75 m^2

22. A triangular prism has a right-triangle base with legs 3 cm and 4 cm. The prism is 12 cm long. The hypotenuse of the triangle is 5 cm. What is the total surface area?

Your Answer:

Find more at
ViewMath.com/MN-Grade7

23. A triangular prism has a base that is a triangle with base 8 cm and height 3 cm. The prism is 10 cm long. What is the volume?

(A) $24\ cm^3$

(B) $80\ cm^3$

(C) $120\ cm^3$

(D) $240\ cm^3$

24. A factory produces 10,000 batteries per day. A sample of 200 batteries is tested, and 8 are defective. What percent of the sample is defective?

Your Answer:

25. The dot plots below show the number of push-ups completed by students in two gym classes.

Class 2

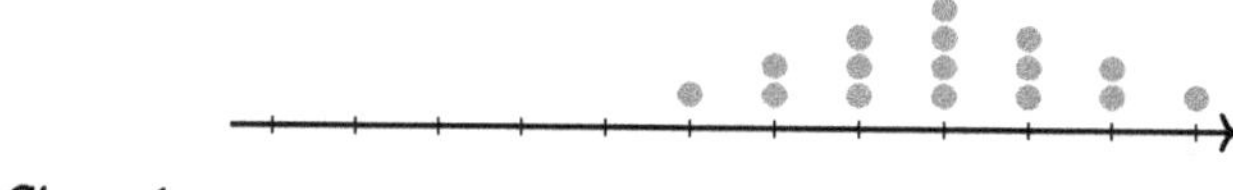

Class 1

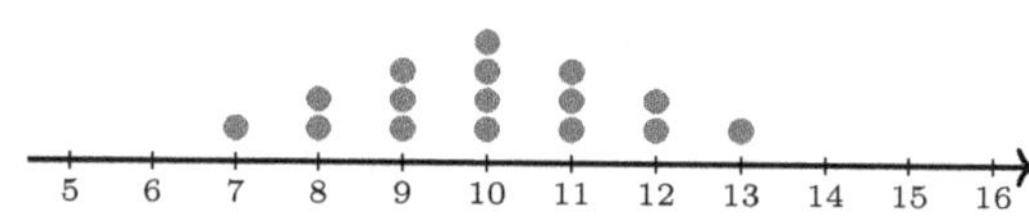

Which class has a higher center, and by approximately how many push-ups?

(A) Class 1, by about 3

(B) Class 2, by about 3

(C) Class 2, by about 6

(D) They have the same center

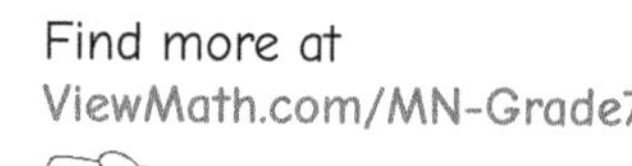

26. *The dot plots show daily high temperatures (°F) for two cities over 10 days.*

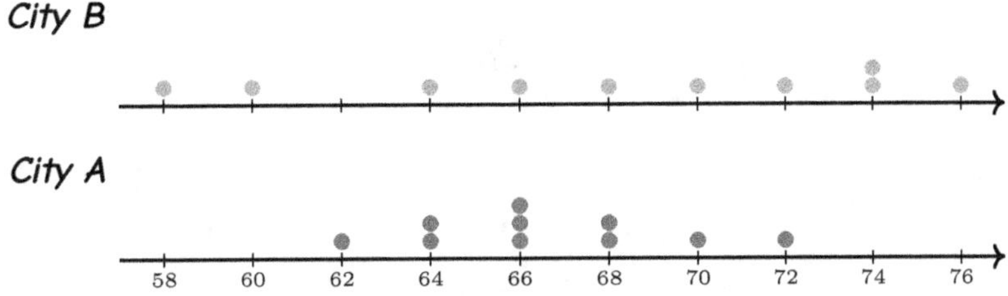

Find the mean and range for each city. Which city has more consistent temperatures?

Your Answer:

27. *A circle graph represents:*

(A) *How data changes over time*

(B) *How a whole is divided into parts*

(C) *The range of a data set*

(D) *Individual data points*

28. *A student creates a probability model for a spinner.* $P(red) = 0.25$, $P(blue) = 0.25$, $P(green) = 0.25$, $P(yellow) = 0.25$. *After 200 spins, the results are: red = 55, blue = 48, green = 52, yellow = 45. Is the model a good fit?*

(A) *No, the model is completely wrong*

(B) *Yes, the observed frequencies are reasonably close to 50 each*

(C) *No, because the results are not exactly 50 each*

(D) *Yes, because 200 is divisible by 4*

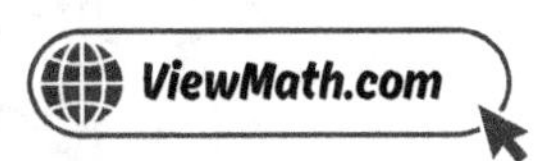

29. The tree diagram shows the outcomes for flipping two coins. What is the probability of getting at least one tail?

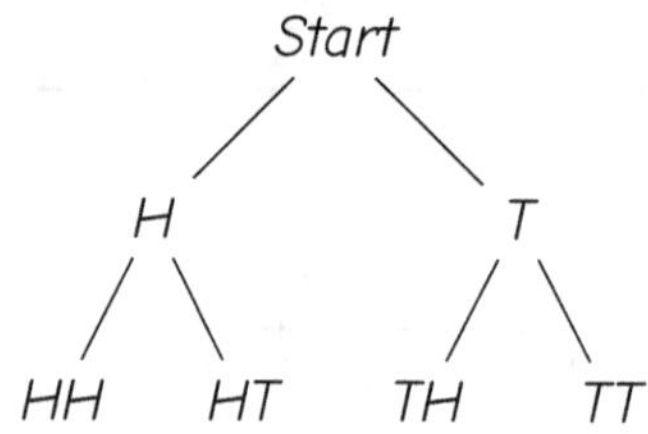

(A) $\frac{1}{4}$

(B) $\frac{1}{2}$

(C) $\frac{3}{4}$

(D) $\frac{2}{4}$

30. The bar graph shows the results of simulating a spinner 100 times. The spinner is supposed to land on each color with equal probability. Based on the simulation, does the spinner appear fair?

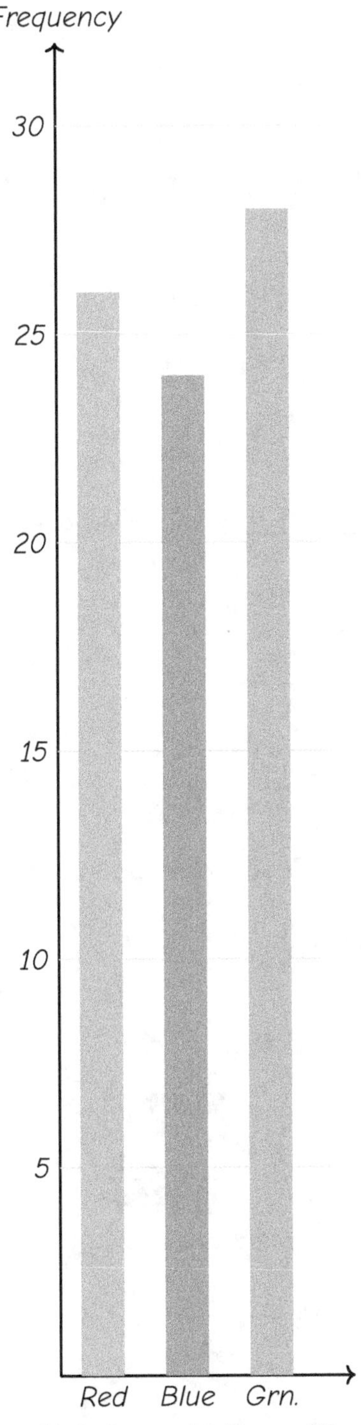

Not shown: Yellow = 22

(A) No, because the bars are different heights

(B) Yes, the frequencies $(26, 24, 28, 22)$ are all roughly close to 25

(C) Green is clearly the most likely color

(D) Cannot be determined from the graph

⭐ End of Practice Test 2 ⭐

Great job finishing the test!

 **My Score**

I got _____________ out of 30 questions right.

*Check your answers in the **Answer Key** at the back of the book.*

Review any questions you missed. That's how we learn!

📊 Check Your Score Online!

Visit **ViewMath Academy** to enter your answers and see which topics you need to review. You can also explore lessons, take quizzes, track your scores, and save your progress!

viewmath.com/score/7.1.MN.17

Or go to viewmath.com/score and enter code: 7.1.MN.17

3

Practice Test 3

 30 Questions

✏️ Before You Start ✏️

- ✓ **Read each question carefully** before choosing your answer.
- ✓ **Show your work** on scratch paper when you need to.
- ✓ **Skip hard questions** and come back to them later.
- ✓ **Check your answers** when you're done.
- ✓ **Take your time** — there's no rush!

 ⭐ You've Got This! ⭐

Do your best and show what you know!

1. *A proportional table has $k = \frac{3}{4}$. Which pair could be in the table?*

(A) $(8, 10)$ (B) $(8, 6)$

(C) $(12, 8)$ (D) $(4, 2)$

2. *Two proportional lines are graphed. Line A is steeper than Line B. What does this tell you?*

(A) Line A has a smaller unit rate (B) Line A has a greater unit rate

(C) Line B covers more distance (D) Both lines have the same rate

3. *Store A sells 3 notebooks for $7.50. Store B sells 5 notebooks for $11.25. Which store offers the better deal?*

(A) Store A at $2.50 each (B) Store B at $2.25 each

(C) Store A at $2.25 each (D) They cost the same

4. *63 is 90% of what number?*

(A) 56.7 (B) 67

(C) 70 (D) 72

5. *Estimate a 15% tip on a $78 bill using the 10% method.*

(A) About $7.80 (B) About $10.40

(C) About $11.70 (D) About $15.60

6. *A student measured a table as 4.5 feet long. The actual length is 5 feet. What is the percent error?*

(A) 0.5% (B) 5%

(C) 10% (D) 11.1%

7. Which is a disadvantage of using a credit card?

   (A) You can buy things before you have the cash

   (B) It builds your credit history

   (C) You pay interest if you don't pay the full balance on time

   (D) It is accepted at most stores

8. Which statement is true?

   (A) $-5 > -2$

   (B) $-3 > 0$

   (C) $-7 < -4$

   (D) $0 < -1$

9. What is the value of $x^2 + 3x$ when $x = -2$?

   (A) 10

   (B) -2

   (C) -10

   (D) 2

10. Simplify $-6y + 4y - y$.

    (A) $-3y$

    (B) $-y$

    (C) $3y$

    (D) $-11y$

11. Solve $9 + 3x = 30$.

    (A) $x = 13$

    (B) $x = 3$

    (C) $x = 7$

    (D) $x = 10$

12. Solve $9(x - 3) = 45$.

    Your Answer:

Find more at
ViewMath.com/MN-Grade7

13. Solve $-0.4y + 2 = 6$.

(A) $y = 10$ (B) $y = -10$

(C) $y = 20$ (D) $y = -20$

14. Solve $-5n + 10 \leq -15$.

Your Answer

15. Solve $-3x + 12 \geq 0$ and describe the graph.

(A) Closed circle at 4, shade right   (B) Open circle at 4, shade left

(C) Closed circle at 4, shade left   (D) Closed circle at -4, shade right

16. A triangle has side lengths 9, 12, and 15. What type of triangle is it?

(A) Equilateral   (B) Isosceles

(C) Right   (D) It cannot be formed

17. The table below shows four sets of triangle conditions. Which one produces a unique triangle?

Set	Given Information	Condition
A	Angles: 50°, 60°, 70°	AAA
B	Sides: 3, 4, 5	SSS
C	Angles: 45°, 90°, 45°	AAA
D	One side: 8 cm	—

(A) Set A   (B) Set B

(C) Set C   (D) Set D

18. You cut a cone with a vertical slice through its apex (tip). What shape is the cross-section?

    (A) Circle

    (B) Triangle

    (C) Rectangle

    (D) Semicircle

19. An angle is 12° less than its complement. What is the angle?

    (A) 39°

    (B) 42°

    (C) 48°

    (D) 51°

20. A circular rug has a diameter of 8 feet. What is the area of the rug? Use $\pi \approx 3.14$.

    (A) 25.12 ft^2

    (B) 50.24 ft^2

    (C) 200.96 ft^2

    (D) 64 ft^2

21. A figure is made of a right triangle with legs 6 cm and 8 cm, and a rectangle 8 cm by 3 cm. What is the total area?

    (A) 24 cm^2

    (B) 48 cm^2

    (C) 36 cm^2

    (D) 72 cm^2

22. The net shown below folds into a rectangular prism. Find the total surface area.

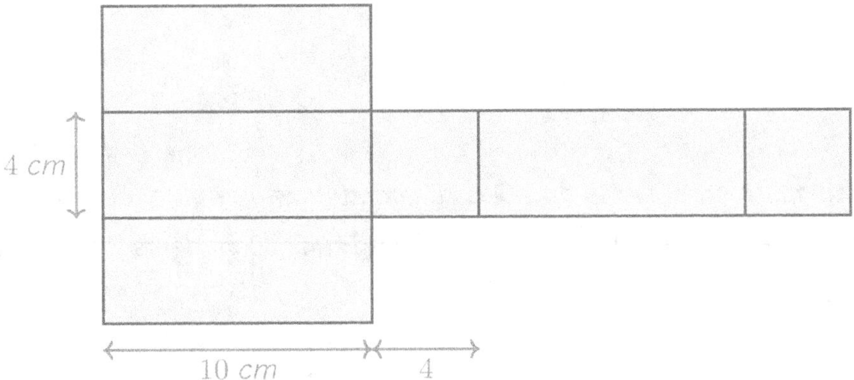

Your Answer:

23. A trapezoidal prism has a trapezoidal base with parallel sides 4 cm and 6 cm, height 3 cm, and the prism length is 10 cm. What is its volume?

(A) $60\ cm^3$

(B) $120\ cm^3$

(C) $150\ cm^3$

(D) $180\ cm^3$

24. Which of the following increases the reliability of conclusions drawn from a sample?

(A) Using a smaller sample

(B) Using a larger random sample

(C) Surveying only people who agree with you

(D) Using a non-random sample

25. School A: median score 88, MAD 4. School B: median score 82, MAD 4. The difference in medians expressed in MADs is:

(A) 1 MAD

(B) 1.5 MADs

(C) 2 MADs

(D) 4 MADs

26. *Group X: mean 45, MAD 6. Group Y: mean 57, MAD 6. Express the difference in means as a multiple of MAD.*

Your Answer

27. *Two circle graphs compare how Class A and Class B spend their free time. Class A shows 45% on sports; Class B shows 30% on sports. If Class A has 20 students and Class B has 30 students, how many more students in Class B play sports than in Class A?*

(A) *0 — the same number*

(B) *9 is more in Class B*

(C) *3 is more in Class A*

(D) *1 more in Class A*

28. *A bag contains 4 red marbles and 1 blue marble. Which type of probability model does this represent?*

(A) *Uniform*

(B) *Non-uniform*

(C) *Both uniform and non-uniform*

(D) *Neither*

29. *Two dice are rolled. What is the probability of getting a sum of 2?*

(A) $\frac{1}{6}$

(B) $\frac{1}{12}$

(C) $\frac{1}{36}$

(D) $\frac{2}{36}$

30. *You simulate flipping a coin 3 times and counting how many times all 3 are heads. In 40 simulations, all 3 heads occurs 6 times. What is the experimental probability?*

(A) $\frac{6}{40} = 0.15$

(B) $\frac{1}{8} = 0.125$

(C) $\frac{3}{40} = 0.075$

(D) $\frac{6}{120} = 0.05$

Find more at
ViewMath.com/MN-Grade7

 End of Practice Test 3 

Great job finishing the test!

☑ My Score

I got _____________ out of 30 questions right.

*Check your answers in the **Answer Key** at the back of the book.*

💡 *Review any questions you missed. That's how we learn!*

📊 Check Your Score Online!

Visit **ViewMath Academy** to enter your answers and see which topics you need to review. You can also explore lessons, take quizzes, track your scores, and save your progress!

viewmath.com/score/7.1.MN.18

Or go to viewmath.com/score and enter code: 7.1.MN.18

Practice Test 4

 30 Questions

✏️ Before You Start ✏️

✓ **Read each question carefully** before choosing your answer.

✓ **Show your work** on scratch paper when you need to.

✓ **Skip hard questions** and come back to them later.

✓ **Check your answers** when you're done.

✓ **Take your time** — there's no rush!

 ⭐ You've Got This! ⭐ 

Do your best and show what you know!

1. What is the constant of proportionality (k) for the table below?

x	y
3	12
5	20
8	32

(A) 3

(B) 4

(C) 8

(D) 12

2. A proportional graph shows total cost versus hours of tutoring. The point $(3, 67.50)$ is on the graph. How much does 5 hours of tutoring cost?

Your Answer:

3. A factory produces 168 items in 7 hours. Another factory produces 200 items in 8 hours. Which factory is more productive?

(A) Factory 1, at 24 items/hr

(B) Factory 2, at 25 items/hr

(C) They produce the same rate

(D) Factory 1, at 25 items/hr

4. A jar has 250 candies. If 44% are chocolate, how many chocolate candies are there?

Your Answer:

5. A restaurant bill is $40. You leave a 15% tip. How much is the tip?

(A) $4.00

(B) $5.00

(C) $6.00

(D) $8.00

Find more at
ViewMath.com/MN-Grade7

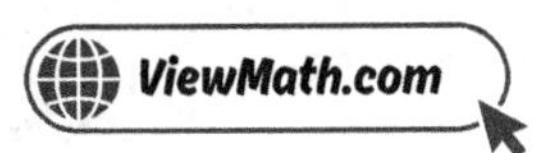

6. In the percent error formula, why do we use absolute value?

  (A) To make the calculation easier   (B) Because the answer must be positive

  (C) Because we only care about the size of the error, not the direction   (D) To convert the answer to a percent

7. Why is it important to pay your credit card bill on time?

  (A) To earn more interest from the credit card company   (B) To avoid paying extra interest charges

  (C) To increase the card's interest rate   (D) To close your bank account

8. Order from least to greatest: $-7,\ 3,\ -1,\ 0,\ -4$

Your Answer:

9. Which expression represents "the product of 7 and a number p, decreased by 12"?

  (A) $7(p - 12)$ (B) $12 - 7p$

  (C) $7p - 12$ (D) $7 + p - 12$

10. Simplify $7x + 3x$.

  (A) $10x$ (B) $21x$

  (C) $10x^2$ (D) $73x$

11. Solve $\dfrac{w}{5} + 12 = 20$.

(A) $w = 40$

(B) $w = 160$

(C) $w = 8$

(D) $w = 1.6$

12. Solve $5(2x - 3) = 35$.

Your Answer:

13. Use the fraction bar model to solve: $\dfrac{2}{3}x = 8$. What is x?

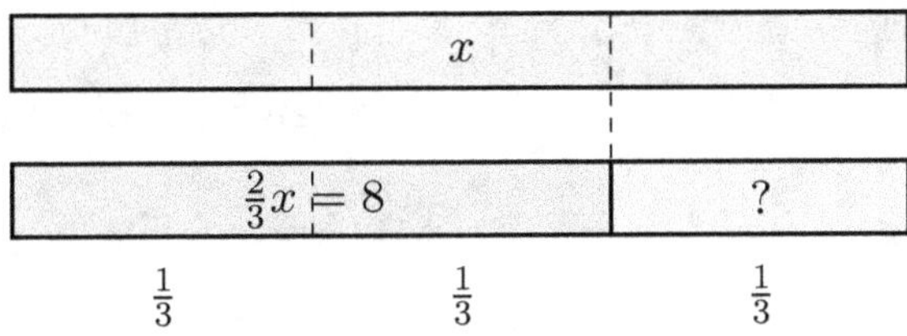

(A) $x = 12$

(B) $x = 16$

(C) $x = 10$

(D) $x = 5\dfrac{1}{3}$

14. Solve $8 - 3x \geq 23$.

Your Answer:

15. Solve $-2x \leq 10$ and describe the graph.

(A) Closed circle at -5, shade left

(B) Closed circle at 5, shade left

(C) Closed circle at -5, shade right

(D) Open circle at -5, shade right

16. The four rectangles below all have the same perimeter of 16 cm but different dimensions. What does this tell us?

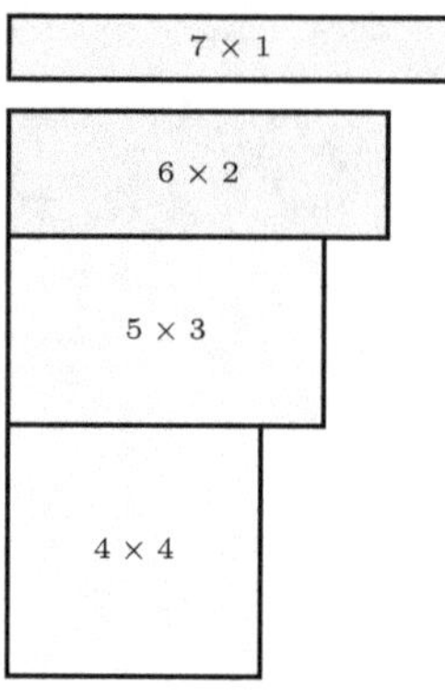

(A) A given perimeter determines exactly one rect-angle

(B) A given perimeter allows more than one rect-angle

(C) All rectangles with the same perimeter have the same area

(D) It is impossible for multiple rectangles to share a perimeter

17. Sides 5, 12, 13. Is this a right triangle?

(A) No, because $5 + 12 \neq 13$

(B) No, because the sides are not equal

(C) Yes, because $5^2 + 12^2 = 13^2$

(D) Yes, because $5 + 12 = 17 > 13$

18. A hexagonal prism is sliced parallel to its base. What is the cross-section?

(A) Rectangle

(B) Triangle

(C) Hexagon

(D) Circle

19. Angles A and B are complementary. Angle A is $(3x - 5)°$ and angle B is $(x + 15)°$. What is angle A?

(A) 20°

(B) 35°

(C) 55°

(D) 65°

20. *What is the area of a circle with a radius of 10 cm? Use $\pi \approx 3.14$.*

(A) $31.4\ cm^2$

(B) $62.8\ cm^2$

(C) $100\ cm^2$

(D) $314\ cm^2$

21. *A shape is made of a square with side 8 cm and a semicircle with diameter 8 cm attached to one side. What is the area? Use $\pi \approx 3.14$.*

(A) $64\ cm^2$

(B) $89.12\ cm^2$

(C) $114.24\ cm^2$

(D) $139.12\ cm^2$

22. *A rectangular prism has dimensions 7 m by 4 m by 3 m. What is the area of the largest face?*

(A) $12\ m^2$

(B) $21\ m^2$

(C) $28\ m^2$

(D) $84\ m^2$

23. *The figure shows unit cubes stacked into an L-shape. Each cube has a side length of 1 cm. What is the total volume?*

(A) $8\ cm^3$

(B) $11\ cm^3$

(C) $15\ cm^3$

(D) $33\ cm^3$

24. *Give one reason why a sample might not perfectly represent a population.*

Your Answer:

25. When comparing two populations visually, you should look at:

(A) Only the center (mean or median)

(B) Only the spread (range or MAD)

(C) Both the center and the spread

(D) Only the largest and smallest values

26. Store A daily sales: mean $500, MAD $40. Store B daily sales: mean $620, MAD $40. The difference in means in MADs is:

(A) 1.5 MADs

(B) 3 MADs

(C) 40 MADs

(D) 120 MADs

27. A student creates a circle graph with sectors of 100°, 80°, 60°, and 100°. Is this circle graph correct?

(A) Yes — the sectors look reasonable

(B) No — the angles add to 340°, not 360°

(C) No — there are too many sectors

(D) Yes — all angles are less than 180°

28. A coin is weighted so that heads comes up 60% of the time. What is the probability of tails?

(A) 0.60

(B) 0.50

(C) 0.40

(D) 0.30

Find more at
ViewMath.com/MN-Grade7

29. Look at the tree diagram for flipping three coins. What is the probability of getting exactly 2 tails? Write your answer as a fraction in simplest form.

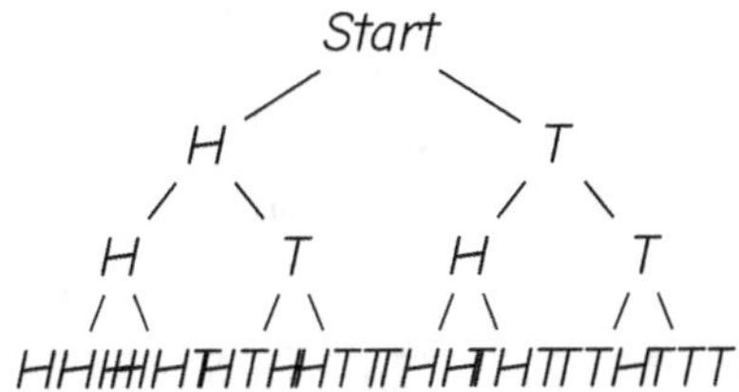

Your Answer:

30. A student wants to simulate the probability that exactly 2 out of 3 students pass a test, where each student has a 50% chance. Which model would work?

(A) Roll one die: odd = pass, even = fail

(B) Flip 3 coins: heads = pass, tails = fail; count exactly 2 heads

(C) Draw 3 marbles from a bag of 2 red and 1 blue

(D) Flip 1 coin 3 times and count all tails

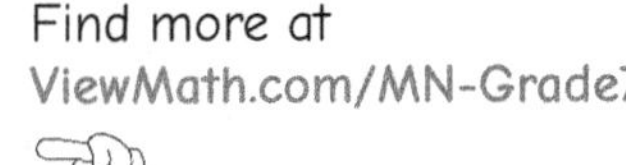

 End of Practice Test 4 

Great job finishing the test!

✅ My Score

I got _____________ out of 30 questions right.

*Check your answers in the **Answer Key** at the back of the book.*

💡 *Review any questions you missed. That's how we learn!*

📊 Check Your Score Online!

Visit **ViewMath Academy** to enter your answers and see which topics you need to review. You can also explore lessons, take quizzes, track your scores, and save your progress!

viewmath.com/score/7.1.MN.19

Or go to viewmath.com/score and enter code: 7.1.MN.19

5

Practice Test 5

 *30 Questions*

 Before You Start 

- ✓ **Read each question carefully** before choosing your answer.
- ✓ **Show your work** on scratch paper when you need to.
- ✓ **Skip hard questions** and come back to them later.
- ✓ **Check your answers** when you're done.
- ✓ **Take your time** — there's no rush!

 You've Got This! 

Do your best and show what you know!

1. A factory produces widgets at a constant rate. In $2\frac{1}{2}$ hours it makes 175 widgets. What is k in widgets per hour?

Your Answer:

2. A proportional graph passes through $(3, 18)$. What is the unit rate?

(A) 3

(B) 6

(C) 15

(D) 18

3. A 12-oz box of cereal costs \$3.60. A 20-oz box costs \$5.40. Which is the better value?

(A) 12-oz box at \$0.30/oz

(B) 20-oz box at \$0.27/oz

(C) 12-oz box at \$0.27/oz

(D) They are the same price per ounce

4. The tape diagram below represents a whole quantity. What percent of the total is the shaded portion?

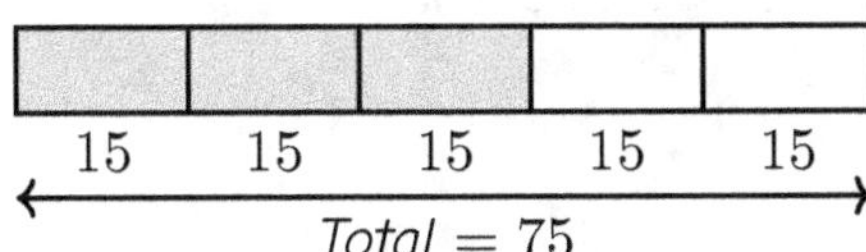

(A) 40%

(B) 45%

(C) 50%

(D) 60%

5. A real estate agent earns 3% commission on a \$250,000 home sale. What does she earn?

(A) \$750

(B) \$2,500

(C) \$7,500

(D) \$75,000

6. An architect estimated a building would be 45 meters tall. The actual height is 42 meters. What is the percent error (rounded to the nearest tenth)?

Your Answer:

7. A credit union is most similar to which of the following?

(A) A department store

(B) A bank

(C) A stock market

(D) A government office

8. Which statement about absolute value is FALSE?

(A) $|-5| = 5$

(B) $|0| = 0$

(C) Absolute value can be negative

(D) $|3| = |-3|$

9. Which expression means "half the sum of a number k and 10"?

(A) $\frac{k}{2} + 10$

(B) $\frac{k+10}{2}$

(C) $\frac{10k}{2}$

(D) $\frac{k}{2+10}$

10. The perimeter of a triangle is $3x + 2 + 5x - 1 + 2x + 4$. Write the simplified expression for the perimeter.

Your Answer:

11. Solve $2x + 15 = 31$.

Your Answer:

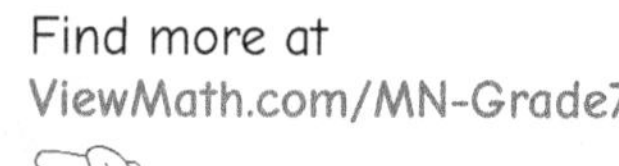

Find more at
ViewMath.com/MN-Grade7

12. Solve $-4(y + 3) = 28$.

(A) $y = 4$

(B) $y = -10$

(C) $y = 10$

(D) $y = -4$

13. Solve $-0.3x + 4.2 = 1.8$.

Your Answer:

14. A movie theater requires you to be more than 48 inches tall for a ride. You are 42 inches tall and grow g inches per year. Which inequality finds how many years until you are tall enough?

(A) $42 + g > 48$

(B) $42g > 48$

(C) $48 - g > 42$

(D) $42 - g > 48$

15. The graph of an inequality has an open circle at -1 and is shaded to the left. Which inequality matches?

(A) $x \leq -1$

(B) $x < -1$

(C) $x \geq -1$

(D) $x > -1$

16. How many different rectangles have an area of 24 cm^2 if the sides must be whole numbers?

Your Answer:

17. Which set of measurements does NOT determine a unique triangle?

(A) Sides 3, 4, 5

(B) Sides 7 and 9 with included angle 50°

(C) Angles 40°, 70°, 70°

(D) Angles 30°, 60° with included side 10 cm

18. You slice a cone with a horizontal cut parallel to its base. What shape is the cross-section?

   (A) A circle smaller than the base

   (B) A circle the same size as the base

   (C) A triangle

   (D) A rectangle

19. Which statement about vertical angles is always true?

   (A) They are supplementary

   (B) They are complementary

   (C) They are equal

   (D) They sum to 360°

20. A sprinkler waters a circular area with a radius of 5 meters. How many square meters does it water? Use $\pi \approx 3.14$.

   (A) $31.4\ m^2$

   (B) $50\ m^2$

   (C) $78.5\ m^2$

   (D) $157\ m^2$

21. An H-shaped figure is made of three rectangles. The top and bottom bars are each 10 cm by 2 cm. The middle vertical bar is 2 cm by 6 cm. What is the total area?

   Your Answer:

22. How many faces does a rectangular prism have?

   (A) 4

   (B) 5

   (C) 6

   (D) 8

23. A moving box is 2 ft long, 2 ft wide, and 3 ft tall. How many of these boxes can fit in a space that is 6 ft by 4 ft by 3 ft?

(A) 3

(B) 6

(C) 12

(D) 18

24. A school has 800 students. The principal surveys 50 students about their favorite subject. What is the population?

(A) The 50 surveyed students

(B) All 800 students in the school

(C) The principal

(D) The students who like math

25. Team A's scores: mean 72, MAD 3. Team B's scores: mean 74, MAD 8. Which team is more consistent?

(A) Team A because its MAD is smaller

(B) Team B because its mean is higher

(C) Team A because its mean is lower

(D) Team B because its MAD is larger

26. Two groups have the same mean but different MADs. What does this tell you?

(A) One group scored higher than the other

(B) The groups have different levels of variability

(C) The groups are identical

(D) The means were calculated incorrectly

27. A sector represents 15% of a data set. What is its angle in degrees?

Your Answer:

Find more at
ViewMath.com/MN-Grade7

28. A bag has tiles labeled A, A, A, B, B. What is the probability of drawing a B?

(A) $\frac{1}{5}$

(B) $\frac{2}{3}$

(C) $\frac{2}{5}$

(D) $\frac{3}{5}$

29. The table shows all possible products when two dice are rolled. How many outcomes give a product of 6? Write the probability as a fraction in simplest form.

×	1	2	3	4	5	6
1	1	2	3	4	5	6
2	2	4	6	8	10	12
3	3	6	9	12	15	18
4	4	8	12	16	20	24
5	5	10	15	20	25	30
6	6	12	18	24	30	36

Your Answer:

30. Which of the following is the FIRST step in designing a simulation?

(A) Record the results

(B) Run many trials

(C) Identify the event and its possible outcomes

(D) Calculate the experimental probability

⭐ End of Practice Test 5 ⭐

Great job finishing the test!

 **My Score**

I got ____________ out of 30 questions right.

*Check your answers in the **Answer Key** at the back of the book.*

💡 *Review any questions you missed. That's how we learn!*

 **Check Your Score Online!**

Visit **ViewMath Academy** to enter your answers and see which topics you need to review. You can also explore lessons, take quizzes, track your scores, and save your progress!

viewmath.com/score/7.1.MN.20

Or go to *viewmath.com/score* and enter code: 7.1.MN.20

Practice Test 6

 30 Questions

 Before You Start

- ✓ **Read each question carefully** before choosing your answer.
- ✓ **Show your work** on scratch paper when you need to.
- ✓ **Skip hard questions** and come back to them later.
- ✓ **Check your answers** when you're done.
- ✓ **Take your time** — there's no rush!

 You've Got This! 

Do your best and show what you know!

1. *A proportional graph passes through $(6, 15)$. What is k?*

Your Answer:

2. *A proportional graph passes through $(4, 10)$. What is y when $x = 7$?*

(A) 13

(B) 17

(C) 17.5

(D) 28

3. *A car uses 3 gallons of gas every 87 miles. How many gallons does it need for a 435-mile trip?*

(A) 10

(B) 12

(C) 15

(D) 18

4. *45 is 75% of what number?*

(A) 33.75

(B) 50

(C) 56

(D) 60

5. *An event planner charges a 10% planning fee plus a 3% coordination fee. If an event costs $4,000, what are the total fees?*

(A) $120

(B) $400

(C) $520

(D) $1,300

6. A scientist predicted a chemical reaction would produce 30 grams. It actually produced 28 grams. What is the percent error?

(A) 2%

(B) 6.7%

(C) 7.1%

(D) 14.3%

7. What does the interest rate on a credit card tell you?

(A) The maximum amount you can borrow

(B) The percent charged for borrowing money

(C) The monthly fee for the card

(D) The amount you must pay each month

8. Which inequality is correct?

(A) $-12 > -3$

(B) $|-4| < |3|$

(C) $-1 > -10$

(D) $|-2| > |5|$

9. Which expression means "a number n divided by 4, then increased by 3"?

(A) $\frac{n+3}{4}$

(B) $\frac{4}{n} + 3$

(C) $\frac{n}{4} + 3$

(D) $4n + 3$

10. Simplify $-4x + 9 + 7x - 3$.

(A) $3x + 6$

(B) $-11x + 6$

(C) $3x - 6$

(D) $11x + 12$

11. Solve $50 - 4x = 18$.

(A) $x = -8$

(B) $x = 17$

(C) $x = 8$

(D) $x = 12$

12. The rectangle below has an area of 56 square cm. What is the value of x?

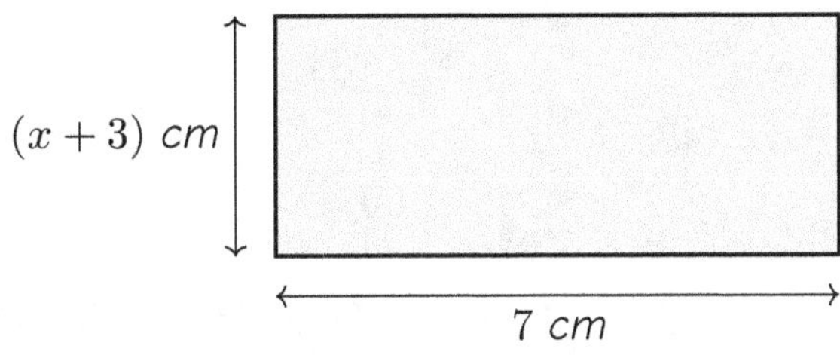

(A) $x = 5$

(B) $x = 8$

(C) $x = 11$

(D) $x = 4$

13. Solve $2.5x - 7.5 = 10$.

Your Answer:

14. A student solved $-2x + 8 > 14$ and wrote $x > -3$. What error did the student make?

(A) Forgot to subtract 8

(B) Divided by 2 instead of -2

(C) Forgot to flip the inequality sign

(D) Added 8 instead of subtracting

15. Solve $5 - x > 8$ and describe the graph.

(A) Open circle at 3, shade right

(B) Open circle at -3, shade left

(C) Closed circle at -3, shade left

(D) Open circle at -3, shade right

16. A circle with radius 5 cm is described as a condition. How many circles satisfy this condition?

(A) None

(B) Exactly one

(C) More than one (different positions)

(D) It depends on the center location

17. Can a triangle have sides 8, 8, and 16? Explain.

Your Answer:

18. Is it possible to get a circular cross-section from a rectangular prism? Explain.

Your Answer:

19. Which pair of angles are supplementary?

(A) 45° and 45°

(B) 60° and 30°

(C) 110° and 70°

(D) 100° and 100°

20. A circle has an area of 12.56 m^2. Using $\pi \approx 3.14$, what is the diameter?

Your Answer:

21. A rectangular room is 14 ft by 10 ft. There is a built-in closet that is 3 ft by 4 ft. What is the area of the room not including the closet?

(A) 128 ft^2

(B) 140 ft^2

(C) 152 ft^2

(D) 12 ft^2

Find more at
ViewMath.com/MN-Grade7

22. A cereal box is 25 cm tall, 18 cm wide, and 5 cm deep. How much cardboard is needed to make the box?

Your Answer:

23. A rectangular prism is 8 in by 5 in by 4 in. Another is 10 in by 4 in by 4 in. Which has more volume and by how much?

(A) First prism, by 10 in^3

(B) Second prism, by 10 in^3

(C) First prism, by 20 in^3

(D) They have the same volume

24. A random sample means that:

(A) Only the best members are chosen

(B) The largest possible group is chosen

(C) Every member of the population has an equal chance of being selected

(D) Only volunteers are selected

25. When is a dot plot a better choice than a box plot for comparing two groups?

(A) When data sets are very large

(B) When you want to see every individual data point

(C) When you only want to compare medians

(D) When the data is categorical

26. Data set: $12, 15, 18, 20, 22, 25, 28$. What is the median?

(A) 18

(B) 20

(C) 22

(D) 19

Find more at
ViewMath.com/MN-Grade7

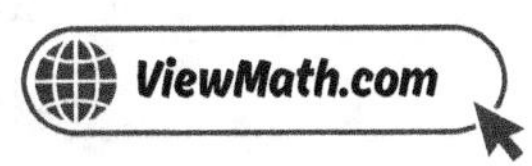

27. A survey of 500 people found: Walking 20%, Driving 55%, Bus 15%, Other 10%. How many drive?

(A) 55

(B) 100

(C) 200

(D) 275

28. A spinner is divided into 4 sections with probabilities: red = 0.4, blue = 0.3, green = 0.2, yellow = 0.1. What type of model is this?

(A) Uniform, because there are 4 sections

(B) Non-uniform, because the probabilities are different

(C) Invalid, because the probabilities don't add up to 1

(D) Uniform, because all sections are on the same spinner

29. A coin is flipped and a die is rolled. What is the probability of getting tails and an even number?

(A) $\frac{1}{12}$

(B) $\frac{1}{6}$

(C) $\frac{1}{4}$

(D) $\frac{1}{2}$

30. What is a simulation?

(A) A way to calculate exact probabilities

(B) A model that uses random numbers or objects to imitate a real event

(C) A method that always gives the same result

(D) A graph that shows all possible outcomes

 End of Practice Test 6 

Great job finishing the test!

☑ My Score

I got __________ out of 30 questions right.

Check your answers in the **Answer Key** at the back of the book.

💡 Review any questions you missed. That's how we learn!

📊 Check Your Score Online!

Visit **ViewMath Academy** to enter your answers and see which topics you need to review. You can also explore lessons, take quizzes, track your scores, and save your progress!

viewmath.com/score/7.1.MN.21

Or go to viewmath.com/score and enter code: 7.1.MN.21

Practice Test 7

 30 Questions

✏ Before You Start ✏

- ✔ **Read each question carefully** before choosing your answer.
- ✔ **Show your work** on scratch paper when you need to.
- ✔ **Skip hard questions** and come back to them later.
- ✔ **Check your answers** when you're done.
- ✔ **Take your time** — there's no rush!

 You've Got This! 

Do your best and show what you know!

1. *Find the missing value in the proportional table.*

x	y
5	20
9	?

Your Answer:

2. *Which statement about proportional graphs is FALSE?*

(A) They always pass through the origin

(B) The slope equals the unit rate

(C) A steeper line means a larger k

(D) The line can be curved

3. *A cyclist rides 42 miles in 3 hours. At this rate, how long does it take to ride 98 miles?*

Your Answer:

4. *9 is 12% of what number?*

(A) 1.08

(B) 36

(C) 72

(D) 75

5. *An online marketplace charges a 4% seller fee. A seller lists an item for \$85. How much does the seller receive after the fee?*

Your Answer:

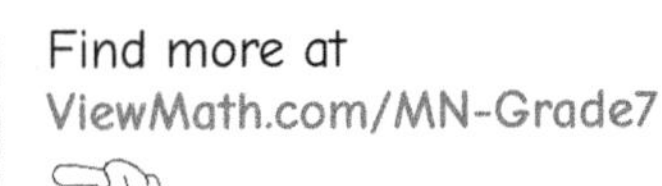
Find more at
ViewMath.com/MN-Grade7

6. You estimated there were 50 marbles in a jar. The actual count was 40. What is the percent error?

(A) 10%

(B) 20%

(C) 25%

(D) 50%

7. A credit card charges 24% annual interest. You owe $500 for 6 months. How much interest do you owe?

Your Answer:

8. Which integer is its own opposite?

(A) 1

(B) −1

(C) 0

(D) No integer is its own opposite

9. Which expression represents "6 more than twice a number n"?

(A) $6n + 2$

(B) $2n + 6$

(C) $2(n + 6)$

(D) $6 \times 2n$

10. *Simplify* $\frac{3}{4}n - \frac{1}{4}n + 5$.

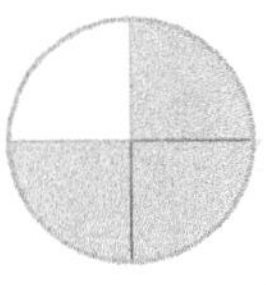

(A) $\frac{1}{2}n + 5$

(B) $n + 5$

(C) $\frac{3}{4}n + 5$

(D) $\frac{2}{4}n - 5$

11. *Solve* $100 - 7x = 37$.

Your Answer

12. *A student solved* $3(x + 4) = 30$ *and wrote:* $3x + 4 = 30$, $3x = 26$, $x = \frac{26}{3}$. *What was the student's error?*

(A) Divided by 3 instead of subtracting

(B) Did not distribute 3 to the 4

(C) Subtracted 4 from the wrong side

(D) Forgot to check the answer

13. *A recipe calls for* $\frac{3}{4}$ *cup of sugar per batch. You already used* $\frac{1}{2}$ *cup. You have* $2\frac{3}{4}$ *cups total. The equation* $\frac{3}{4}b + \frac{1}{2} = 2\frac{3}{4}$ *finds how many more batches b you can make. What is b?*

(A) $b = 3$

(B) $b = 4$

(C) $b = 2$

(D) $b = 3\frac{2}{3}$

14. Which inequality means "no more than 25"?

   (A) $x > 25$

   (B) $x \geq 25$

   (C) $x < 25$

   (D) $x \leq 25$

15. Solve $3x + 4 > 16$. Describe the graph on a number line.

   Your Answer:

16. Can you draw a triangle with sides 5 cm, 5 cm, and 5 cm?

   (A) No, it is impossible

   (B) Yes, exactly one triangle

   (C) Yes, more than one triangle

   (D) Yes, but only if one angle is 90°

17. A triangle has sides 6 cm, 8 cm, and 10 cm. How many triangles can be constructed with these measurements?

   (A) None

   (B) Exactly one

   (C) Exactly two

   (D) Infinitely many

18. Which statement about cross-sections is true?

   (A) The cross-section of any 3D figure is always a circle

   (B) The cross-section depends on the angle and position of the cut

   (C) A cross-section is always the same shape as the base

   (D) Only prisms can have cross-sections

19. An angle measures 38°. Its complement measures:

   (A) 38°

   (B) 52°

   (C) 142°

   (D) 322°

Find more at
ViewMath.com/MN-Grade7

20. Find the area of a semicircle with a diameter of 10 cm. Use $\pi \approx 3.14$.

Your Answer:

21. A T-shaped figure can be broken into a 12 cm by 2 cm horizontal rectangle and a 2 cm by 8 cm vertical rectangle. What is the total area?

(A) $24 \ cm^2$

(B) $32 \ cm^2$

(C) $40 \ cm^2$

(D) $48 \ cm^2$

22. What does surface area measure?

(A) The space inside a 3D figure

(B) The distance around a 3D figure

(C) The total area of all faces of a 3D figure

(D) The height of a 3D figure

23. Find the volume of a rectangular prism with dimensions 9 cm by 5 cm by 4 cm.

Your Answer:

24. The diagram below shows four different sampling methods used to survey students at a school. Which method is most likely to produce an unbiased sample?

Method A

Survey the football team

Method B

Survey every 10th student on the school roster

Method C

Post survey online and wait for responses

Method D

Survey students in one math class

(A) Method A

(B) Method B

(C) Method C

(D) Method D

25. Class A: median 90, IQR 6. Class B: median 83, IQR 14. Which class is more consistent?

Your Answer

26. Class A: median 85, IQR 8, range 20. Class B: median 80, IQR 15, range 40. A student says "Class B is better because it has a wider range." Is the student correct?

(A) Yes — wider range means more talent

(B) No — wider range means less consistency, and Class A has a higher median

(C) Yes — Class B's scores spread more, which is always better

(D) No — range does not measure anything useful

27. A survey of 120 students asked about their favorite after-school activity. The results are:

Activity	Students
Sports	48
Music	30
Art	18
Reading	24

Find the percent and angle (in degrees) for each category.

Your Answer:

28. Which of the following is an example of a uniform probability model?

(A) A bag with 3 red, 5 blue, and 2 green marbles

(B) A fair number cube with 6 sides

(C) A spinner with sections of different sizes

(D) A weighted coin

29. A coin is flipped and a die is rolled. What is the probability of getting heads and a number greater than 4?

(A) $\frac{1}{12}$

(B) $\frac{1}{6}$

(C) $\frac{2}{12}$

(D) $\frac{1}{3}$

30. A game has a 25% chance of winning. Which simulation model correctly represents this?

(A) Flip a coin — heads means win

(B) Roll a die — roll a 1 means win

(C) Use a spinner with 4 equal sections — one section means win

(D) Use digits 0–9 — digits 0–4 mean win

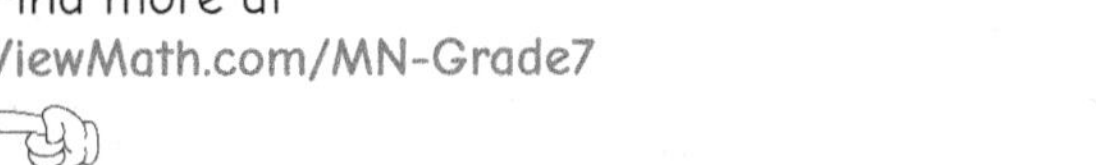

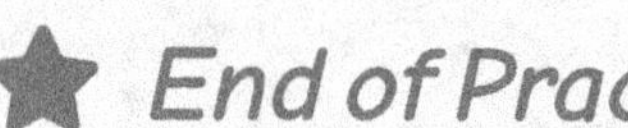 End of Practice Test 7 

Great job finishing the test!

✅ My Score

I got _____________ out of 30 questions right.

*Check your answers in the **Answer Key** at the back of the book.*

💡 *Review any questions you missed. That's how we learn!*

📊 Check Your Score Online!

Visit **ViewMath Academy** to enter your answers and see which topics you need to review. You can also explore lessons, take quizzes, track your scores, and save your progress!

viewmath.com/score/7.1.MN.22

Or go to viewmath.com/score and enter code: 7.1.MN.22

Practice Test 8

 30 Questions

 ✏️ Before You Start ✏️ 

- ✔ **Read each question carefully** before choosing your answer.
- ✔ **Show your work** on scratch paper when you need to.
- ✔ **Skip hard questions** and come back to them later.
- ✔ **Check your answers** when you're done.
- ✔ **Take your time** — there's no rush!

 ⭐ You've Got This! ⭐ 

Do your best and show what you know!

1. A car travels at a constant speed. It goes 150 miles in 2.5 hours. What is the constant of proportionality?

(A) 30 mph

(B) 60 mph

(C) 75 mph

(D) 150 mph

2. Two friends start walking at the same time. The graph shows their distances. How many miles apart are they after 6 hours?

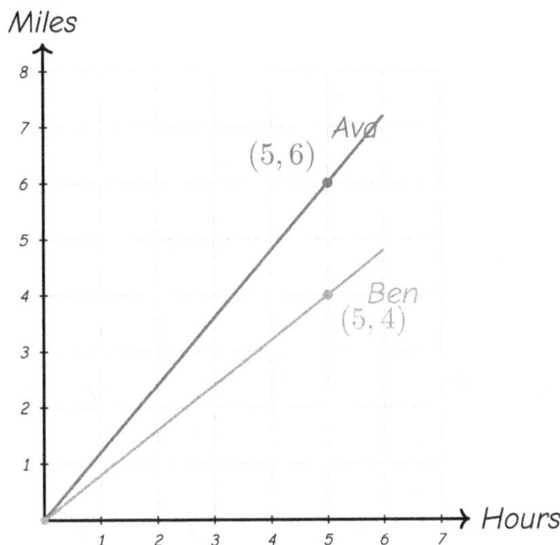

Your Answer:

3. A scale drawing uses 1 cm = 4.5 m. If a building is 27 m tall, how tall is it in the drawing?

Your Answer:

4. What is 150% of 40?

(A) 20

(B) 40

(C) 50

(D) 60

5. *Estimate a 20% tip on a $93 bill. Round the bill first, then calculate.*

Your Answer

6. *The number line below shows an estimated value and an actual value. What is the percent error?*

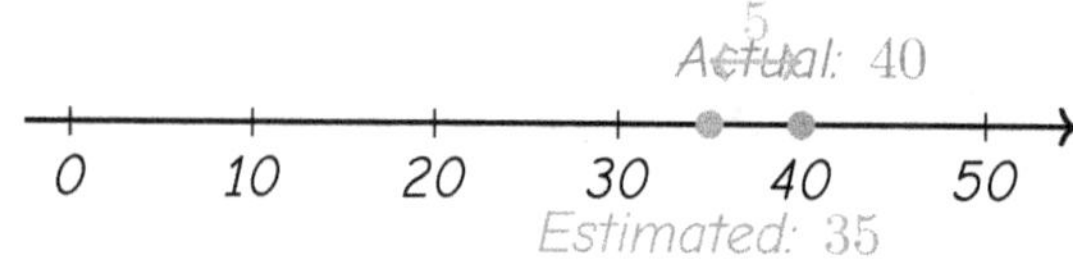

(A) 5%

(B) 10%

(C) 12.5%

(D) 14.3%

7. *You charge $400 on a credit card at 10% annual interest. You pay $200 right away and owe $200 for 1 year. How much total interest do you pay?*

(A) $40

(B) $20

(C) $10

(D) $60

8. *For which value of n is $|n| = 10$?*

(A) *Only $n = 10$*

(B) *Only $n = -10$*

(C) *$n = 10$ or $n = -10$*

(D) *$n = 0$*

9. *Evaluate $-4y + 10$ when $y = 3$.*

(A) 22

(B) -2

(C) 2

(D) -22

10. Write an expression with three terms that simplifies to $4n + 7$.

Your Answer:

11. The table shows a pattern. What is the value of x when the output is 41?

Input (x)	Output
1	8
2	11
3	14
?	41

A) $x = 12$

B) $x = 11$

C) $x = 13$

D) $x = 14$

12. Solve $5(n + 3) = -10$.

A) $n = -5$

B) $n = 1$

C) $n = -1$

D) $n = -\dfrac{25}{5}$

13. Solve $\dfrac{5n}{6} + \dfrac{1}{3} = 3$.

Your Answer:

14. The table shows ticket prices for groups. A school has \$200 to spend and must also pay a \$35 bus fee. Write and solve an inequality to find the maximum number of students *s* who can go.

Group Size	Price per Student
1–10	\$12
11–25	\$9
26+	\$7

Assume the group qualifies for the \$9 rate.

Your Answer:

15. Solve $10 - 4x < 2$. Describe the graph.

Your Answer:

16. The shortest two sides of a triangle are 7 and 9. What is the largest whole-number length the third side can be?

Your Answer:

17. What condition (SSS, SAS, ASA, AAS, or AAA) gives infinitely many triangles?

Your Answer:

18. Which 3D figure always has circular cross-sections when cut parallel to its base?

(A) Rectangular prism

(B) Triangular pyramid

(C) Cone

(D) Cube

Find more at
ViewMath.com/MN-Grade7

ViewMath.com

19. Two supplementary angles are in the ratio 5 : 4. Find both angles.

Your Answer:

20. A square has a circle inscribed inside it as shown. The side of the square is 10 cm. What is the area of the circle? Use $\pi \approx 3.14$

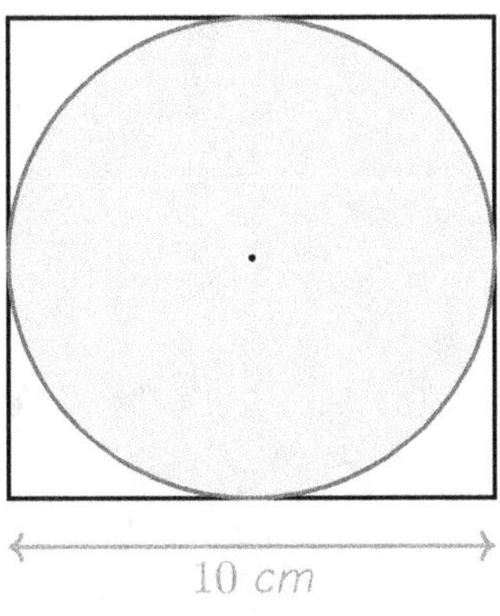

(A) 31.4 cm^2

(B) 78.5 cm^2

(C) 100 cm^2

(D) 314 cm^2

21. A rectangular patio is 12 ft by 8 ft. A circular hot tub with a diameter of 6 ft is placed on the patio. What is the area of the patio NOT covered by the hot tub? Use $\pi \approx 3.14$

Your Answer:

22. What is the surface area of a rectangular prism with length 5 cm, width 3 cm, and height 2 cm?

(A) 30 cm^2

(B) 62 cm^2

(C) 46 cm^2

(D) 31 cm^2

23. What is the volume of a cube with side length 5 m?

(A) $15\ m^3$

(B) $25\ m^3$

(C) $125\ m^3$

(D) $150\ m^3$

24. A random sample of 50 apples from an orchard found that 6 were bruised. If the orchard has 3,000 apples, about how many are bruised?

Your Answer:

25. Team A: mean 60, MAD 4. Team B: mean 70, MAD 3. Is the difference in means meaningful? Explain briefly.

Your Answer:

26. Data set: $20, 22, 25, 27, 30, 32, 35$. What is the IQR?

(A) 15

(B) 10

(C) 7

(D) 5

27. A survey of 60 people: 18 chose option A, 24 chose option B, 18 chose option C. Find the angle for option B in a circle graph.

Your Answer:

Find more at
ViewMath.com/MN-Grade7

28. A bag contains 5 yellow and 5 purple marbles. Lisa picks a marble, records the color, and replaces it 40 times. She gets yellow 24 times. Which statement is correct?

(A) The model predicts $P(yellow) = 0.5$, and the experiment gives $P(yellow) = 0.6$, so there is a small difference

(B) The model is wrong because $24 \neq 20$

(C) The model predicts $P(yellow) = 0.6$

(D) The experiment proves yellow is more likely than purple

29. A coin is flipped and a die is rolled. What is the probability of getting tails and a 1? Write your answer as a fraction.

Your Answer

30. How does increasing the number of trials in a simulation affect the results?

(A) The results become less reliable

(B) The experimental probability gets closer to the theoretical probability

(C) The results stay exactly the same

(D) The simulation takes less time

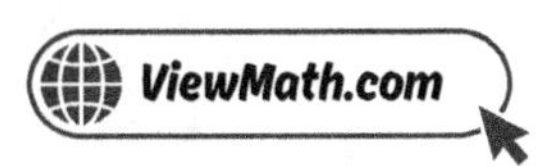

 End of Practice Test 8 

Great job finishing the test!

My Score

I got ___________ out of 30 questions right.

*Check your answers in the **Answer Key** at the back of the book.*

Review any questions you missed. That's how we learn!

Check Your Score Online!

Visit **ViewMath Academy** to enter your answers and see which topics you need to review. You can also explore lessons, take quizzes, track your scores, and save your progress!

viewmath.com/score/7.1.MN.23

Or go to *viewmath.com/score* and enter code: 7.1.MN.23

Practice Test 9

 *30 Questions*

 Before You Start 

- ✓ **Read each question carefully** before choosing your answer.
- ✓ **Show your work** on scratch paper when you need to.
- ✓ **Skip hard questions** and come back to them later.
- ✓ **Check your answers** when you're done.
- ✓ **Take your time** — there's no rush!

 You've Got This! 

Do your best and show what you know!

1. Two students find k from the point $(8, 20)$. Student A says $k = 2.5$. Student B says $k = 0.4$. Who is correct?

   (A) Student A, because $k = \frac{y}{x}$

   (B) Student B, because $k = \frac{x}{y}$

   (C) Both are correct, depending on which variable is independent

   (D) Neither is correct

2. The equation $y = 5x$ gives the cost of movie tickets. What does the point $(1, 5)$ represent?

   (A) 5 movies cost \$1

   (B) 1 ticket costs \$5

   (C) The total for all tickets is \$5

   (D) You need 5 hours to buy 1 ticket

3. A printer prints 120 pages in 8 minutes. How long will it take to print 450 pages?

   (A) 25 min

   (B) 28 min

   (C) 30 min

   (D) 36 min

4. 14 is what percent of 56?

   (A) 14%

   (B) 20%

   (C) 25%

   (D) 28%

5. A hair stylist charges \$55 for a haircut. The customer leaves a 20% tip. What is the total the customer pays?

   (A) \$11.00

   (B) \$60.50

   (C) \$66.00

   (D) \$75.00

Find more at
ViewMath.com/MN-Grade7

6. A builder estimated a project would cost $8,000. The actual cost was $7,500. What is the percent error?

  (A) 5%   (B) 6.25%

  (C) 6.67%   (D) 10%

7. Bank A pays 1.5% interest on savings. Bank B pays 4%. You deposit $2,000 for 1 year. How much more does Bank B earn you?

  (A) $25   (B) $50

  (C) $80   (D) $5

8. A vertical number line shows the elevations of four locations.

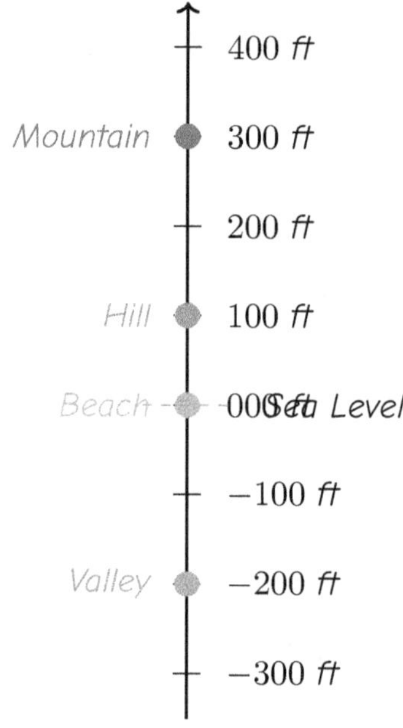

What is the difference in elevation between the Mountain and the Valley?

Your Answer

9. Which phrase best describes the expression $5n - 8$?

(A) 8 less than 5 times a number

(B) 5 less than 8 times a number

(C) 8 minus 5 times a number

(D) 5 times a number less 8 times that number

10. A rectangular garden has the dimensions shown. Write a simplified expression for the total distance around the garden (its perimeter).

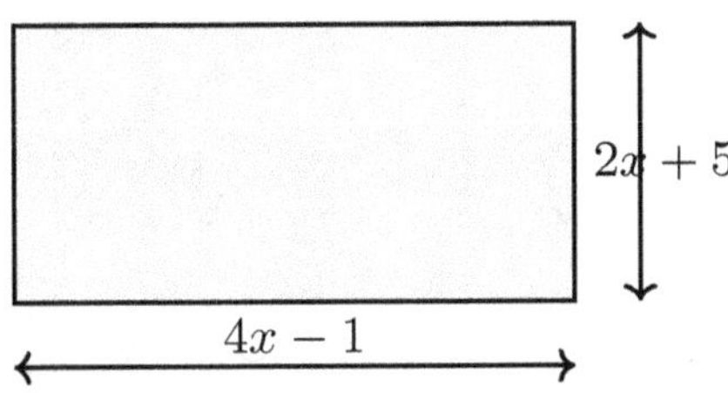

Your Answer:

11. Which equation has the solution $x = 5$?

(A) $4x + 3 = 18$

(B) $4x + 3 = 23$

(C) $4x - 3 = 23$

(D) $4x - 5 = 25$

12. Solve $2(x + 6) = 20$.

(A) $x = 4$

(B) $x = 7$

(C) $x = 10$

(D) $x = 14$

13. A shirt is on sale for $\frac{2}{3}$ of its original price. After using a $5 coupon, you pay $15. What equation finds the original price p?

(A) $\frac{2}{3}p + 5 = 15$

(B) $p - \frac{2}{3} - 5 = 15$

(C) $\frac{2}{3}p - 5 = 15$

(D) $\frac{2}{3}(p - 5) = 15$

14. Solve $-3n + 6 \leq 0$.

(A) $n \leq 2$

(B) $n \leq -2$

(C) $n \geq 2$

(D) $n \geq -2$

15. Solve $\frac{x}{2} - 3 > 1$ and describe the graph.

(A) Open circle at 8, shade right

(B) Closed circle at 8, shade right

(C) Open circle at 4, shade right

(D) Open circle at 8, shade left

16. A square must have an area of 36 cm^2. How many different squares satisfy this condition?

(A) None

(B) Exactly one

(C) Exactly four

(D) More than one

17. Can a triangle have sides 4, 4, and 9?

(A) Yes, it is an isosceles triangle

(B) Yes, it is a scalene triangle

(C) No, the triangle inequality fails

(D) No, because two sides are equal

18. You slice a sphere with any flat plane. What shape is the cross-section?

   (A) Oval

   (B) Rectangle

   (C) Circle

   (D) It depends on the angle of the cut

19. An angle is 4 times its supplement. What is the angle?

   (A) 36°

   (B) 45°

   (C) 120°

   (D) 144°

20. A quarter-circle has a radius of 8 cm. What is its area? Use $\pi \approx 3.14$.

   (A) 200.96 cm^2

   (B) 100.48 cm^2

   (C) 50.24 cm^2

   (D) 25.12 cm^2

21. A pentagon can be divided into a rectangle 6 cm by 4 cm and a triangle with base 6 cm and height 2 cm. What is the total area?

   (A) 24 cm^2

   (B) 30 cm^2

   (C) 36 cm^2

   (D) 48 cm^2

22. What is the surface area of a cube with side length 4 cm?

   (A) 16 cm^2

   (B) 64 cm^2

   (C) 96 cm^2

   (D) 24 cm^2

Find more at
ViewMath.com/MN-Grade7

23. A rectangular prism has a base area of 24 cm^2 and a height of 7 cm. What is its volume?

(A) 31 cm^3

(B) 168 cm^3

(C) 336 cm^3

(D) 17 cm^3

24. A researcher wants to know the average height of seventh graders in a state. She measures the heights of 100 seventh graders from different schools. What is the sample?

(A) All seventh graders in the state

(B) All students in the schools she visited

(C) The 100 seventh graders she measured

(D) The tallest seventh graders

25. The box plots below compare test scores for Class A and Class B.

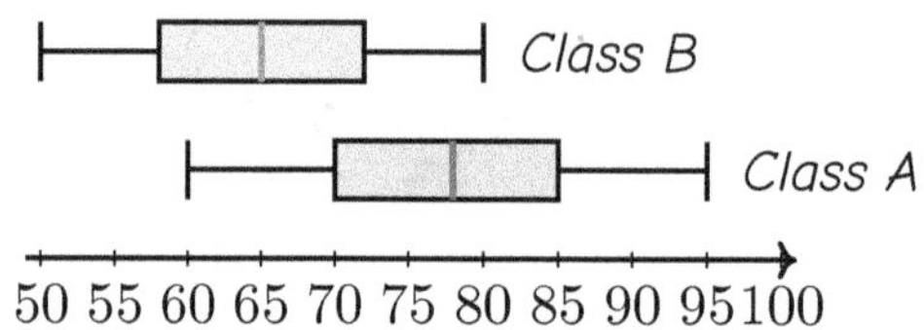

Which statement is best supported by the box plots?

(A) Class B scored higher than Class A

(B) Class A has a higher median and a larger IQR

(C) Class A has a higher median and both classes have the same IQR

(D) Class A and Class B overlap, but Class A's median is higher

26. Group A: $\{60, 65, 70, 75, 80\}$. Group B: $\{40, 55, 70, 85, 100\}$. Both have a mean of 70. Which group is more variable?

(A) Group A

(B) Group B

(C) They are equally variable

(D) Cannot be determined

Find more at
ViewMath.com/MN-Grade7

27. The circle graph shows how a student spends a 24-hour day.

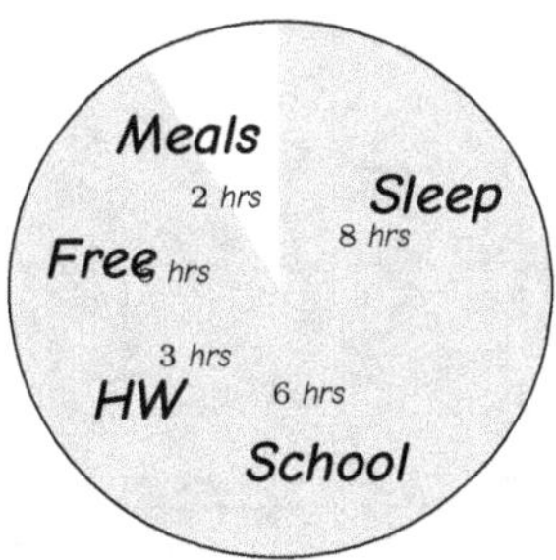

What percent of the day does the student spend on homework?

(A) 3%

(B) 8.3%

(C) 12.5%

(D) 25%

28. A teacher predicted that a spinner would land on red $\frac{1}{4}$ of the time. After 80 spins, red appeared 25 times. How do the observed and predicted results compare?

(A) They are very different — the model is wrong

(B) The observed $(\frac{25}{80} = 0.3125)$ is close to the predicted (0.25)

(C) The observed result is exactly equal to the predicted result

(D) The predicted probability is higher than the observed

29. Three coins are flipped. What is the probability of getting NO heads (all tails)?

(A) $\frac{1}{2}$

(B) $\frac{1}{4}$

(C) $\frac{1}{8}$

(D) $\frac{3}{8}$

30. Describe how to use a standard die to simulate a $\frac{1}{3}$ probability event.

Your Answer:

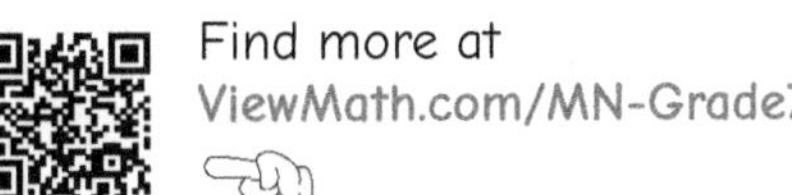
Find more at
ViewMath.com/MN-Grade7

 # End of Practice Test 9 

Great job finishing the test!

 My Score

I got ___________ out of 30 questions right.

Check your answers in the **Answer Key** at the back of the book.

Review any questions you missed. That's how we learn!

📊 Check Your Score Online!

Visit **ViewMath Academy** to enter your answers and see which topics you need to review. You can also explore lessons, take quizzes, track your scores, and save your progress!

viewmath.com/score/7.1.MN.24

Or go to viewmath.com/score and enter code: 7.1.MN.24

10

Practice Test 10

30 Questions

✏ Before You Start ✏

- ✓ **Read each question carefully** before choosing your answer.
- ✓ **Show your work** on scratch paper when you need to.
- ✓ **Skip hard questions** and come back to them later.
- ✓ **Check your answers** when you're done.
- ✓ **Take your time** — there's no rush!

★ You've Got This! ★

Do your best and show what you know!

1. The graph below shows a proportional relationship. What is the constant of proportionality?

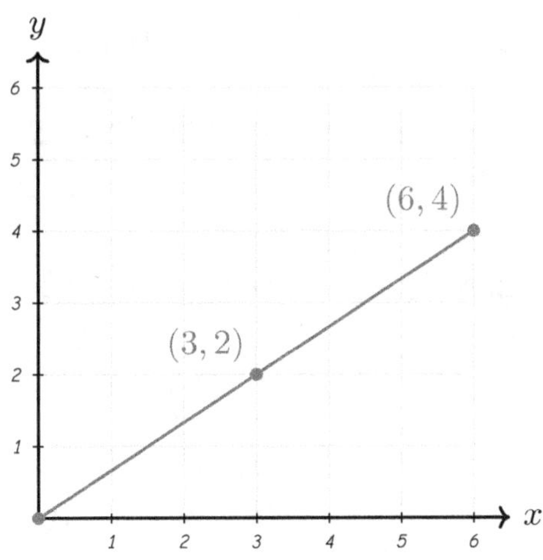

(A) $k = \frac{1}{2}$

(B) $k = \frac{2}{3}$

(C) $k = \frac{3}{2}$

(D) $k = 2$

2. The graph below shows a proportional relationship between hours and miles traveled. What is the unit rate?

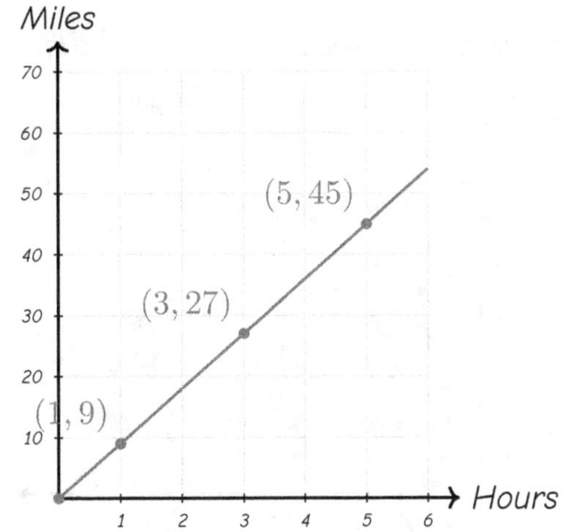

(A) 3 mph

(B) 9 mph

(C) 27 mph

(D) 45 mph

3. A map scale says 2 cm = 15 miles. Two cities are 7 cm apart on the map. What is the actual distance?

(A) 42.5 miles

(B) 45 miles

(C) 52.5 miles

(D) 105 miles

4. A library has 480 books. If 35% are fiction, how many fiction books does the library have?

(A) 148

(B) 160

(C) 168

(D) 172

5. Two friends split a $70 dinner bill equally and each leave a 15% tip on their share. How much does each person pay in total?

(A) $35.00

(B) $37.25

(C) $40.25

(D) $45.50

6. The bar graph shows the estimated and actual amounts of water (in liters) collected during an experiment over three days. On which day was the percent error the largest, and what was it?

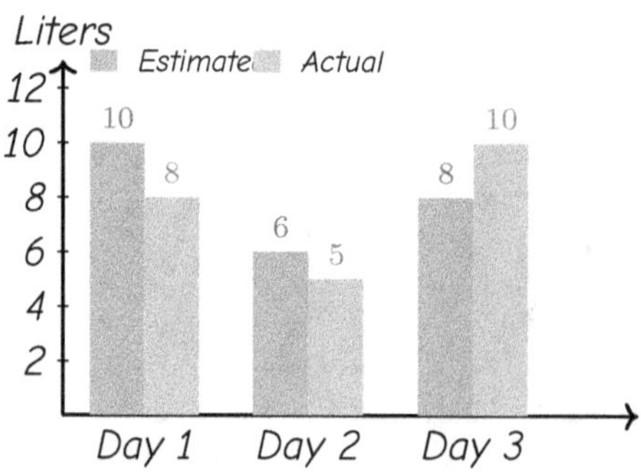

Your Answer:

7. Bank A has a \$4 monthly fee. Bank B has no monthly fee. How much do you save per year at Bank B?

Your Answer:

8. A submarine is at -200 meters. It rises 200 meters. What is its new depth?

Your Answer:

9. Evaluate $2a + 7$ when $a = -3$.

(A) 13

(B) 1

(C) -13

(D) -1

10. Simplify $3y + 8 - 5y - 2$.

(A) $8y + 6$

(B) $-2y + 6$

(C) $2y - 6$

(D) $-2y + 10$

11. The bar model shows an equation. Solve for x.

45			
x	x	x	6

Your Answer:

12. *Solve* $2(4x - 3) = 26$.

(A) $x = 3$ (B) $x = 4$

(C) $x = 2$ (D) $x = 5$

13. *The bar graph shows how much three students earned. Together they earned a total of $31.50. Student C earned $\frac{1}{2}$ as much as Student A. Find how much Student C earned.*

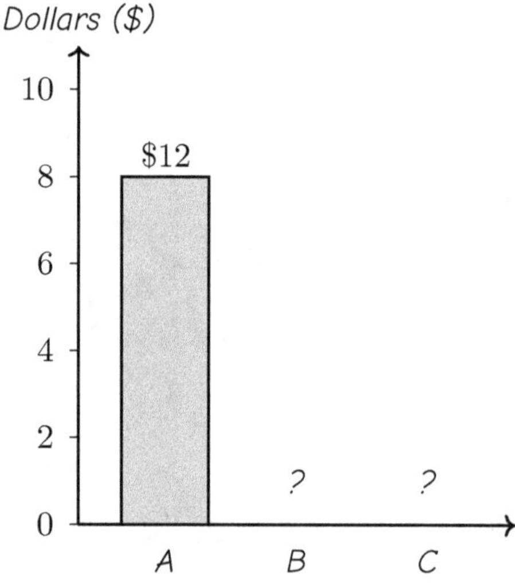

Student A earned $12, and Student C earned $\frac{1}{2}$ of Student A's earnings.

Your Answer:

14. *Solve* $-x + 9 \leq 15$.

(A) $x \leq -6$ (B) $x \geq 6$

(C) $x \geq -6$ (D) $x \leq 6$

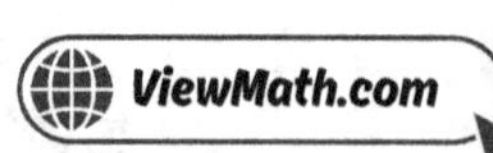

15. Solve $6 - 2x \leq -4$ and describe the graph.

(A) Closed circle at 5, shade right

(B) Open circle at 5, shade right

(C) Closed circle at 5, shade left

(D) Closed circle at −5, shade right

16. Can a triangle have sides 10 cm, 6 cm, and 3 cm? Explain why or why not.

Your Answer:

17. Can a triangle have angles 90°, 100°, and −10°?

(A) Yes, it is a right triangle

(B) Yes, because the sum is 180°

(C) No, because an angle cannot be negative

(D) No, because the sum is not 180°

18. The diagram shows a cylinder with radius 4 cm and height 9 cm. A vertical cut is made through the center. Describe the cross-section and give its dimensions.

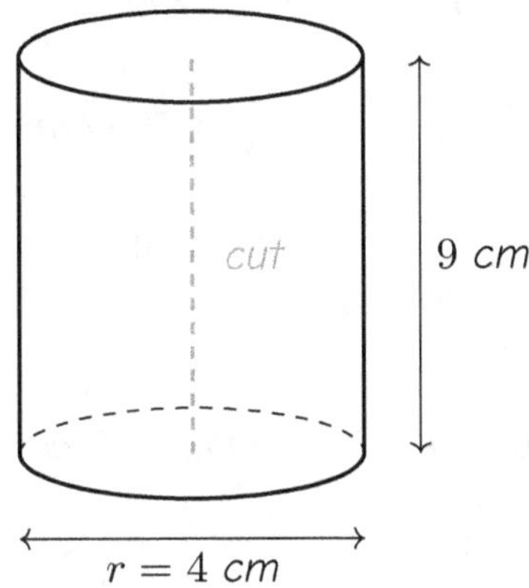

Your Answer:

Find more at
ViewMath.com/MN-Grade7

19. Two angles are supplementary. One is 25° more than the other. What are the two angles?

(A) 65° and 115°

(B) 77.5° and 102.5°

(C) 80° and 100°

(D) 75° and 105°

20. A pizza has a radius of 7 inches. What is the area of the pizza? Use $\pi \approx \frac{22}{7}$.

(A) 44 in^2

(B) 154 in^2

(C) 308 in^2

(D) 22 in^2

21. A composite shape is made of a rectangle that is 10 cm by 4 cm and a triangle with base 10 cm and height 3 cm attached to one side. What is the total area?

(A) 40 cm^2

(B) 55 cm^2

(C) 70 cm^2

(D) 120 cm^2

22. A cereal box is 30 cm tall, 20 cm wide, and 6 cm deep. What is its surface area?

(A) 1,440 cm^2

(B) 1,560 cm^2

(C) 1,800 cm^2

(D) 3,600 cm^2

23. A rectangular prism has a volume of 360 cm^3. Its length is 12 cm and its height is 5 cm. What is its width?

Your Answer

Find more at
ViewMath.com/MN-Grade7

24. The table below shows the results of two samples taken from a population of 1,000 students.

	Sample 1 (50 students)	Sample 2 (50 students)
Prefer Math	18	22
Prefer Science	14	12
Prefer English	10	8
Prefer Art	8	8

Based on both samples, the best prediction for how many of the 1,000 students prefer math is:

(A) 180

(B) 220

(C) 400

(D) 360

25. Two histograms show that Group A's data is skewed left and Group B's data is symmetric. Which measure of center is most appropriate to compare them?

(A) Mean for both

(B) Median for both

(C) Mode for both

(D) Range for both

26. Data: $100, 200, 300, 400, 500$. Find the median and the IQR.

Your Answer:

27. The full circle in a circle graph represents:

(A) 50%

(B) 100% or 360°

(C) 180°

(D) The largest category

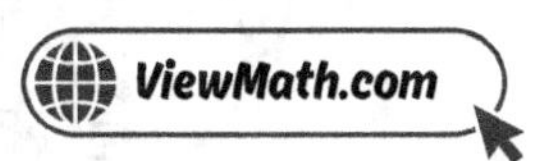

Find more at
ViewMath.com/MN-Grade7

28. Look at the spinner below. Which probability model correctly represents this spinner?

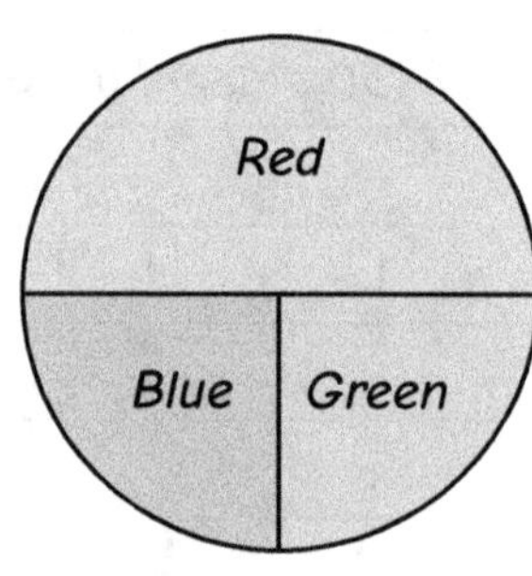

(A) $P(Red) = \frac{1}{3}$, $P(Blue) = \frac{1}{3}$, $P(Green) = \frac{1}{3}$

(B) $P(Red) = \frac{1}{2}$, $P(Blue) = \frac{1}{4}$, $P(Green) = \frac{1}{4}$

(C) $P(Red) = \frac{1}{4}$, $P(Blue) = \frac{1}{4}$, $P(Green) = \frac{1}{2}$

(D) $P(Red) = \frac{1}{2}$, $P(Blue) = \frac{1}{2}$, $P(Green) = 0$

29. Two dice are rolled. What is the probability that the product (not the sum) of the two numbers is 12?

(A) $\frac{2}{36}$

(B) $\frac{4}{36}$

(C) $\frac{3}{36}$

(D) $\frac{6}{36}$

30. A student simulates rolling two dice 36 times to see how often the sum is 7. They get a sum of 7 a total of 8 times. How does this compare to the expected number?

(A) Expected: 6; the simulation result (8) is slightly higher

(B) Expected: 6; the simulation result (8) is much higher

(C) Expected: 12; the simulation result (8) is lower

(D) Expected: 7; the simulation result (8) is exact

 *End of Practice Test 10* 

Great job finishing the test!

☑ My Score

I got ___________ out of 30 questions right.

*Check your answers in the **Answer Key** at the back of the book.*

💡 *Review any questions you missed. That's how we learn!*

📊 Check Your Score Online!

*Visit **ViewMath Academy** to enter your answers and see which topics you need to review. You can also explore lessons, take quizzes, track your scores, and save your progress!*

viewmath.com/score/7.1.MN.25

Or go to viewmath.com/score and enter code: 7.1.MN.25

Answer Key & Explanations

Answer Key

First try each test on your own, then check your work here.

✓ Practice Test 1 — Answer Key

1 $k = 2.5$; missing $x = 6$	**2** B	**3** A	**4** C
5 B	**6** $\approx 4.2\%$	**7** D	**8** D
9 6	**10** $\frac{5}{4}x - 2$ or $1\frac{1}{4}x - 2$	**11** A	**12** $t = \$14$
13 $p = \$60$	**14** A	**15** $x \geq -6$	
16 B	**17** Yes	**18** A triangle	**19** B
20 B	**21** B	**22** A	**23** C
24 B	**25** C	**26** B	**27** C
28 0.4	**29** B	**30** 48	

💡 Time to Learn! 💡

*Review the explanations below, **especially for the questions you missed**.*

Understanding why each answer is correct builds stronger problem-solving skills.

***Tip:** Circle any questions you got wrong, then read their explanation carefully.*

📖 Practice Test 1 — Detailed Explanations

1 $k = \frac{7.5}{3} = 2.5$. For the missing row: $15 = 2.5x$ so $x = 15 \div 2.5 = 6$.

Find more at
ViewMath.com/MN-Grade7

2. Line P: $k = \frac{7}{5} = 1.4$. Line Q: $k = \frac{6}{6} = 1$. Line P is steeper and has the greater unit rate.

3. Worker A: $60 \div 5 = 12$ boxes/hr. Worker B: $45 \div 5 = 9$ boxes/hr. Worker A is faster by $12 - 9 = 3$ boxes/hr.

4. $60\% = 0.60$. Multiply: $0.60 \times 300 = 180$ items.

5. Commission: $6{,}000 \times 0.05 = \$300$. Total: $400 + 300 = \$700$.

6. $\frac{|5-4.8|}{4.8} \times 100 = \frac{0.2}{4.8} \times 100 \approx 4.17 \approx 4.2\%$.

7. Online Bank charges $\$0$/month, so the annual fee is $\$0$. That is the lowest.

8. $8 + (-8) = 0$. Gaining and losing the same amount are opposite actions that cancel out.

9. Substitute: $\frac{5^2-1}{4} = \frac{25-1}{4} = \frac{24}{4} = 6$.

10. Find a common denominator: $\frac{2}{4}x + \frac{3}{4}x = \frac{5}{4}x$. The constant stays: $\frac{5}{4}x - 2$.

11. Perimeter: $(2x+1) + (x+3) + (x+5) = 39$. Combine: $4x + 9 = 39$. Subtract 9: $4x = 30$. Divide by 4: $x = 7.5$.

12. $4(t+3) = 68$. Divide by 4: $t + 3 = 17$. Subtract 3: $t = 14$. Each ticket costs $\$14$.

13. After $\frac{1}{4}$ markdown: $p - \frac{p}{4} = \frac{3p}{4}$. After subtracting $\$8$: $\frac{3p}{4} - 8 = 37$. Add 8: $\frac{3p}{4} = 45$. Multiply by $\frac{4}{3}$: $p = 60$.

14. Subtract 3: $2x > 8$. Divide by 2: $x > 4$.

15. Closed circle means -6 is included, shading right means values greater than or equal to -6: $x \geq -6$.

16 Two sides and the included angle (SAS) fully determine one unique triangle.

17 $11 + 13 = 24 > 20$ ✓, $11 + 20 = 31 > 13$ ✓, $13 + 20 = 33 > 11$ ✓. All three checks pass.

18 A vertical slice through the apex of a cone produces an isosceles triangle. The base of the triangle is a diameter of the cone's base.

19 A straight line measures 180°. Adjacent angles that together form a straight line must sum to 180° (they are supplementary).

20 $r = 7$ cm. $A = \frac{22}{7} \times 49 = 154$ cm^2.

21 Rectangle: $12 \times 5 = 60$ cm^2. Triangle: $\frac{1}{2} \times 4 \times 5 = 10$ cm^2. Remaining: $60 - 10 = 50$ cm^2.

22 Two triangles: $2 \times \frac{1}{2} \times 6 \times 4 = 24$ cm^2. Three rectangles: $6 \times 10 + 5 \times 10 + 5 \times 10 = 60 + 50 + 50 = 160$ cm^2. Total: $24 + 160 = 184$ cm^2.

23 $V = 50 \times 30 \times 40 = 60{,}000$ cm^3.

24 $\frac{15}{60} = \frac{1}{4} = 25\%$. So 25% of $600 = 150$.

25 A box plot shows the quartiles (Q_1 and Q_3), so you can read the IQR directly as $Q_3 - Q_1$.

26 Company X has smaller IQR (1.5 vs. 4) and range (4 vs. 8), meaning delivery times are more predictable.

27 $\frac{16}{40} = 0.40 = 40\%$.

28 $P(Z) = 1 - 0.15 - 0.45 = 0.40$.

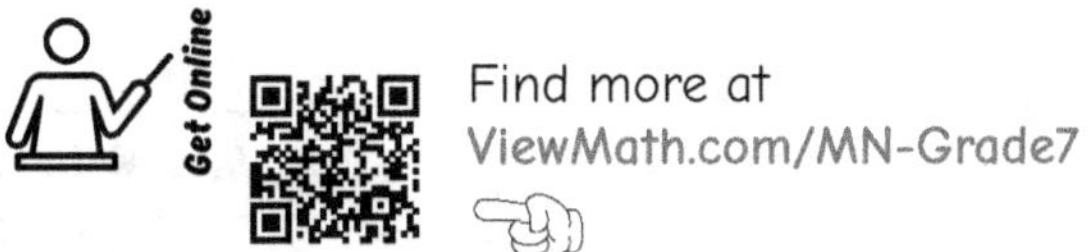

29 $P(spinner = 3) = \frac{1}{4}$, $P(die = 3) = \frac{1}{6}$. $P = \frac{1}{4} \times \frac{1}{6} = \frac{1}{24}$.

30 $80 \times 0.60 = 48$.

✅ Practice Test 2 — Answer Key

1 B   2 0.75 birdhouses per hour; 7.5 birdhouses in 10 hours   3 $37\frac{1}{2}$ cups   4 B   5 D

6 C   7 C   8 7 and -7 9 C   10 $5a + 2b$ 11 B   12 $x = -10$ 13 $x = 14$

14 $x \le 10$ 15 D   16 D   17 B   18 B   19 137°   20 B   21 B   22 $156\ cm^2$

23 C   24 4%   25 B

26 City A: mean = 66.6°F, range = 10°F. City B: mean = 68.2°F, range = 18°F. City A is more consistent.

27 B   28 B   29 C   30 B

💡 Time to Learn! 💡

Review the explanations below, **especially for the questions you missed.**

Understanding why each answer is correct builds stronger problem-solving skills.

Tip: Circle any questions you got wrong, then read their explanation carefully.

📖 Practice Test 2 — Detailed Explanations

1 $k = \frac{54}{2} = 27$, $\frac{135}{5} = 27$, $\frac{216}{8} = 27$. The car gets 27 miles per gallon.

Find more at
ViewMath.com/MN-Grade7

2 $k = \frac{3}{4} = 0.75$. In 10 hours: $0.75 \times 10 = 7.5$ birdhouses.

3 $k = \frac{7.5}{3} = 2.5$ cups per batch. For 15 batches: $2.5 \times 15 = 37.5 = 37\frac{1}{2}$ cups.

4 $75\% = 0.75$. Multiply: $0.75 \times 32 = 24$ students passed.

5 $17.20 \div 86 = 0.20 = 20\%$.

6 A: $\frac{|180-200|}{200} = 10\%$. B: $\frac{|220-200|}{200} = 10\%$. Both have 10% error.

7 Interest $= 300 \times 0.18 \times 1 = \$54$.

8 $|7| = 7$ and $|-7| = 7$. A number and its opposite share the same absolute value.

9 Substitute: $\frac{10}{2} - 4 = 5 - 4 = 1$.

10 Combine a-terms: $-2a + 7a = 5a$. Combine b-terms: $5b - 3b = 2b$. Result: $5a + 2b$.

11 Subtract 2: $6 = \frac{x}{3}$. Multiply by 3: $x = 18$. Check: $\frac{18}{3} + 2 = 6 + 2 = 8$ ✓.

12 Divide by 2: $x + 7 = -3$. Subtract 7: $x = -10$. Check: $2(-10 + 7) = 2(-3) = -6$ ✓.

13 Subtract 1.5: $\frac{x}{4} = 3.5$. Multiply by 4: $x = 14$. The point is at 14 on the number line.

14 Subtract 7: $\frac{x}{2} \leq 5$. Multiply by 2: $x \leq 10$.

15 Closed circle means the endpoint is included, shading left means less than or equal to: $x \leq 7$.

16 Three angles (AAA) determine the shape but not the size. You can draw infinitely many similar triangles of different sizes with these angles.

17 $180° − 35° − 75° = 70°.$

18 A diagonal plane that passes through all six faces of a cube creates a regular hexagonal cross-section.

19 $180° − 43° = 137°.$

20 $r^2 = A \div \pi = 78.5 \div 3.14 = 25.$ So $r = 5$ cm.

21 Rectangle: $25 \times 10 = 250$ m^2. Semicircle: $\frac{1}{2} \times 3.14 \times 5^2 = \frac{1}{2} \times 78.5 = 39.25$ m^2. Total: $250 + 39.25 = 289.25$ m^2.

22 Two triangles: $2 \times \frac{1}{2} \times 3 \times 4 = 12$ cm^2. Three rectangles: $3 \times 12 + 4 \times 12 + 5 \times 12 = 36 + 48 + 60 = 144$ cm^2. Total: $12 + 144 = 156$ cm^2.

23 $B = \frac{1}{2} \times 8 \times 3 = 12$ cm^2. $V = Bh = 12 \times 10 = 120$ cm^3.

24 $\frac{8}{200} = 0.04 = 4\%.$

25 Class 1 center ≈ 10. Class 2 center ≈ 13. Difference ≈ 3 push-ups. Class 2 is higher.

26 City A: $(62 + 64 + 64 + 66 + 66 + 66 + 68 + 68 + 70 + 72) \div 10 = 666 \div 10 = 66.6$. Range $= 72 − 62 = 10$. City B: $(58 + 60 + 64 + 66 + 68 + 70 + 72 + 74 + 74 + 76) \div 10 = 682 \div 10 = 68.2$. Range $= 76 − 58 = 18$. City A has the smaller range.

27 A circle graph (pie chart) shows how a whole quantity is divided into categories.

Find more at
ViewMath.com/MN-Grade7

28. Each color should appear about 50 times (200×0.25). The results $(55, 48, 52, 45)$ are all close to 50, supporting the model.

29. Outcomes with at least one tail: HT, TH, TT — that is 3 out of 4. $P = \frac{3}{4}$.

30. With 4 equal sections, each should appear about 25 times in 100 spins. The results $(26, 24, 28, 22)$ are all close to 25, suggesting the spinner is fair.

✅ Practice Test 3 — Answer Key

1 B	2 B	3 B	4 C	5 C	6 C	7 C	8 C	9 B	10 A
11 C	12 $x = 8$	13 B	14 $n \geq 5$	15 C	16 C	17 B	18 B	19 A	
20 B	21 B	22 $192 \ cm^2$	23 C	24 B	25 B	26 2 MADs	27 A	28 B	
29 C	30 A								

💡 Time to Learn! 💡

Review the explanations below, **especially for the questions you missed**.

Understanding why each answer is correct builds stronger problem-solving skills.

Tip: Circle any questions you got wrong, then read their explanation carefully.

📖 Practice Test 3 — Detailed Explanations

1. If $k = \frac{3}{4}$, then $y = \frac{3}{4}x$. For $x = 8$: $y = \frac{3}{4} \times 8 = 6$. So $(8, 6)$ works.

Find more at
ViewMath.com/MN-Grade7

2 A steeper line means a higher slope, which equals a greater unit rate (constant of proportionality).

3 Store A: $7.50 \div 3 = \$2.50$ each. Store B: $11.25 \div 5 = \$2.25$ each. Store B is cheaper.

4 Divide: $63 \div 0.90 = 70$.

5 10% of $78 = \$7.80$. Half of that (5%) $= \$3.90$. Add: $7.80 + 3.90 = \$11.70$.

6 $\frac{|4.5-5|}{5} \times 100 = \frac{0.5}{5} \times 100 = 10\%$.

7 The main disadvantage is paying extra money (interest) if you don't pay the full balance by the due date.

8 On a number line, -7 is to the left of -4, so $-7 < -4$.

9 Substitute: $(-2)^2 + 3(-2) = 4 + (-6) = -2$.

10 $-6y + 4y = -2y$. Then $-2y - y = -3y$.

11 Subtract 9: $3x = 21$. Divide by 3: $x = 7$. Check: $9 + 3(7) = 9 + 21 = 30$ ✓.

12 Divide by 9: $x - 3 = 5$. Add 3: $x = 8$. Check: $9(8 - 3) = 9(5) = 45$ ✓.

13 Subtract 2: $-0.4y = 4$. Divide by -0.4: $y = -10$. Check: $-0.4(-10) + 2 = 4 + 2 = 6$ ✓.

14 Subtract 10: $-5n \leq -25$. Divide by -5 and flip: $n \geq 5$.

15 Subtract 12: $-3x \geq -12$. Divide by -3 and flip: $x \leq 4$. Closed circle at 4, shade left.

Find more at
ViewMath.com/MN-Grade7

16 Check: $9^2 + 12^2 = 81 + 144 = 225 = 15^2$. Since the Pythagorean theorem holds, this is a right triangle.

17 SSS with valid side lengths ($3 + 4 = 7 > 5$, etc.) produces exactly one unique triangle. AAA gives many triangles, and one side alone is insufficient.

18 A vertical cut through the apex of a cone produces a triangle (specifically an isosceles triangle).

19 Let the angle be x and its complement be $x + 12$. Sum: $x + (x + 12) = 90$. Combine: $2x + 12 = 90$. Solve: $2x = 78$, $x = 39°$.

20 $r = 8 \div 2 = 4$ ft. $A = \pi r^2 = 3.14 \times 16 = 50.24$ ft^2.

21 Triangle: $\frac{1}{2} \times 6 \times 8 = 24$ cm^2. Rectangle: $8 \times 3 = 24$ cm^2. Total: $24 + 24 = 48$ cm^2.

22 The prism is $10 \times 4 \times 4$. $SA = 2(10 \times 4) + 2(10 \times 4) + 2(4 \times 4) = 80 + 80 + 32 = 192$ cm^2.

23 $B = \frac{1}{2}(4 + 6) \times 3 = 15$ cm^2. $V = 15 \times 10 = 150$ cm^3.

24 Larger random samples better represent the population and give more reliable results.

25 Difference in medians: $88 - 82 = 6$. In MADs: $\frac{6}{4} = 1.5$ MADs.

26 Difference: $57 - 45 = 12$. In MADs: $\frac{12}{6} = 2$.

27 Class A: 45% of 20 = 9. Class B: 30% of 30 = 9. Same number: 9 each.

28 The probability of drawing red ($\frac{4}{5}$) is different from drawing blue ($\frac{1}{5}$), so this is a non-uniform model.

29. The only way to get a sum of 2 is $(1, 1)$. $P = \frac{1}{36}$.

30. $P = \frac{6}{40} = 0.15$. The theoretical probability is $\frac{1}{8} = 0.125$, close to the simulation result.

✅ Practice Test 4 — Answer Key

| 1 | B | 2 | $112.50 | 3 | B | 4 | 110 | 5 | C | 6 | C | 7 | B | 8 | $-7, -4, -1, 0, 3$ |

9. C   10. A   11. A   12. $x = 5$ 13. A   14. $x \leq -5$ 15. C   16. B   17. C

18. C   19. C   20. D   21. B   22. C   23. B

24. Random variation (or the sample may be biased)   25. C   26. B   27. B   28. C   29. $\frac{3}{8}$

30. B

💡 Time to Learn! 💡

Review the explanations below, **especially for the questions you missed**.

Understanding why each answer is correct builds stronger problem-solving skills.

Tip: Circle any questions you got wrong, then read their explanation carefully.

📖 Practice Test 4 — Detailed Explanations

1. $k = \frac{y}{x} = \frac{12}{3} = 4$. Check: $\frac{20}{5} = 4$ and $\frac{32}{8} = 4$

2. $k = \frac{67.50}{3} = 22.50$ *per hour. Cost for 5 hours:* $22.50 \times 5 = \$112.50.$

3. *Factory 1:* $168 \div 7 = 24$ *items/hr. Factory 2:* $200 \div 8 = 25$ *items/hr. Factory 2 is more productive.*

4. $44\% = 0.44.$ *Multiply:* $0.44 \times 250 = 110$ *chocolate candies.*

5. $15\% = 0.15.$ *Tip:* $40 \times 0.15 = \$6.00.$

6. *Absolute value ensures the percent error is positive whether the estimate was too high or too low.*

7. *Paying on time avoids interest charges. Late payments mean you pay more for the same purchase.*

8. *On a number line from left to right:* $-7 < -4 < -1 < 0 < 3.$

9. *"The product of 7 and p" is $7p$. "Decreased by 12" means subtract 12:* $7p - 12.$

10. $7x$ *and* $3x$ *are like terms. Add the coefficients:* $7 + 3 = 10.$ *Result:* $10x.$

11. *Subtract 12:* $\frac{w}{5} = 8.$ *Multiply by 5:* $w = 40.$ *Check:* $\frac{40}{5} + 12 = 8 + 12 = 20$ ✓.

12. *Divide by 5:* $2x - 3 = 7.$ *Add 3:* $2x = 10.$ *Divide by 2:* $x = 5.$ *Check:* $5(2(5) - 3) = 5(7) = 35$ ✓.

13. *If* $\frac{2}{3}x = 8$, *then each* $\frac{1}{3}$ *of x is 4. So* $x = 3 \times 4 = 12.$

14. *Subtract 8:* $-3x \geq 15.$ *Divide by -3 and flip:* $x \leq -5.$

15. *Divide by -2 and flip:* $x \geq -5.$ *Closed circle at -5, shade right.*

16 The four rectangles all have perimeter 16 cm ($2(7+1) = 2(6+2) = 2(5+3) = 2(4+4) = 16$) but different shapes. A given perimeter allows many rectangles.

17 $5^2 + 12^2 = 25 + 144 = 169 = 13^2$. The Pythagorean theorem holds, so it is a right triangle.

18 Any cut parallel to the base of a prism gives a shape congruent to the base. A hexagonal prism has a hexagonal base, so the cross-section is a hexagon.

19 $3x - 5 + x + 15 = 90$. Combine: $4x + 10 = 90$. Solve: $4x = 80$, $x = 20$. Angle A: $3(20) - 5 = 55°$.

20 $A = \pi r^2 = 3.14 \times 100 = 314$ cm^2.

21 Square: $8^2 = 64$ cm^2. Semicircle: $\frac{1}{2} \times \pi r^2 = \frac{1}{2} \times 3.14 \times 16 = 25.12$ cm^2. Total: $64 + 25.12 = 89.12$ cm^2.

22 The three pairs of faces have areas $7 \times 4 = 28$, $7 \times 3 = 21$, and $4 \times 3 = 12$ m^2. The largest is 28 m^2.

23 Bottom row: 5 cubes. Second row: 3 cubes. Third row: 3 cubes. Total: $5 + 3 + 3 = 11$ cm^3.

24 Even a well-chosen random sample may differ slightly from the population due to chance.

25 A complete comparison requires looking at both center (where the data clusters) and spread (how spread out it is).

26 Difference $= \$620 - \$500 = \$120$. In MADs: $\frac{120}{40} = 3$ MADs.

27 $100 + 80 + 60 + 100 = 340°$, which is $20°$ short of $360°$. The graph is incorrect.

28 $P(tails) = 1 - 0.60 = 0.40$. Since heads and tails have different probabilities, this is a non-uniform model.

29 Outcomes with exactly 2 tails: HTT, THT, TTH — that is 3 out of 8. $P = \frac{3}{8}$.

30 Each coin flip has a 50% chance of heads (pass). Flipping 3 coins and counting exactly 2 heads simulates exactly 2 passes.

✅ Practice Test 5 — Answer Key

1 $k = 70$ widgets per hour	2 B	3 B	4 D	5 C	6 $\approx 7.1\%$	7 B

8 C   9 B   10 $10x + 5$ 11 $x = 8$ 12 B   13 $x = 8$ 14 A   15 B   16 4

17 C   18 A   19 C   20 C   21 52 cm^2 22 C   23 B   24 B   25 A

26 B   27 54°   28 C   29 $\frac{1}{9}$ 30 C

💡 Time to Learn! 💡

Review the explanations below, **especially for the questions you missed**.

Understanding why each answer is correct builds stronger problem-solving skills.

Tip: Circle any questions you got wrong, then read their explanation carefully.

📖 Practice Test 5 — Detailed Explanations

1 $k = \frac{175}{2.5} = 70$ widgets per hour.

2 Unit rate $= k = \frac{18}{3} = 6$.

Find more at
ViewMath.com/MN-Grade7

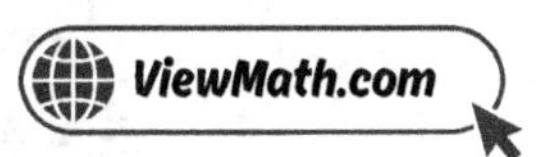

3. 12-oz: $3.60 \div 12 = \$0.30/oz$. 20-oz: $5.40 \div 20 = \$0.27/oz$. The 20-oz box is the better value.

4. The shaded portion is $3 \times 15 = 45$. The total is 75. Percent: $45 \div 75 = 0.60 = 60\%$.

5. $250{,}000 \times 0.03 = \$7{,}500$.

6. $\frac{|45-42|}{42} \times 100 = \frac{3}{42} \times 100 \approx 7.14 \approx 7.1\%$.

7. A credit union works like a bank (savings, checking, loans) but is owned by its members and often has lower fees.

8. Absolute value measures distance from 0, which is always ≥ 0. It can never be negative.

9. "The sum of k and 10" is $k + 10$. "Half" of that sum is $\frac{k+10}{2}$.

10. Combine variable terms: $3x + 5x + 2x = 10x$. Combine constants: $2 - 1 + 4 = 5$. Perimeter: $10x + 5$.

11. Subtract 15: $2x = 16$. Divide by 2: $x = 8$. Check: $2(8) + 15 = 31$ ✓.

12. Divide by -4: $y + 3 = -7$. Subtract 3: $y = -10$. Check: $-4(-10 + 3) = -4(-7) = 28$ ✓.

13. Subtract 4.2: $-0.3x = -2.4$. Divide by -0.3: $x = 8$. Check: $-0.3(8) + 4.2 = -2.4 + 4.2 = 1.8$ ✓.

14. Current height plus growth: $42 + g > 48$.

15. Open circle means -1 is not included, shading left means less than: $x < -1$.

16. Factor pairs of 24: 1×24, 2×12, 3×8, 4×6. That is 4 different rectangles.

Find more at
ViewMath.com/MN-Grade7

17 AAA alone does not fix the size of the triangle. Many triangles can share these angle measures.

18 A horizontal cut through a cone parallel to the base produces a circle smaller than the base. The higher you cut, the smaller the circle.

19 Vertical angles are always equal. They are formed by two intersecting lines, and the opposite angles are congruent.

20 $A = \pi r^2 = 3.14 \times 25 = 78.5 \ m^2$.

21 Top: $10 \times 2 = 20 \ cm^2$. Bottom: $10 \times 2 = 20 \ cm^2$. Middle: $2 \times 6 = 12 \ cm^2$. Total: $20 + 20 + 12 = 52 \ cm^2$.

22 A rectangular prism has 6 faces: top, bottom, front, back, left, and right.

23 Space volume: $6 \times 4 \times 3 = 72 \ ft^3$. Box volume: $2 \times 2 \times 3 = 12 \ ft^3$. Number: $72 \div 12 = 6$ boxes.

24 The population is the entire group you want to study — all 800 students.

25 A smaller MAD means scores are closer to the mean, so Team A is more consistent.

26 Same mean means same center. Different MADs mean one group's data is more spread out than the other's.

27 $0.15 \times 360° = 54°$.

28 There are 2 B tiles out of 5 total. $P(B) = \frac{2}{5}$.

29 Products of 6: $(1,6),(2,3),(3,2),(6,1)$ — 4 outcomes. $P = \frac{4}{36} = \frac{1}{9}$.

30 Before choosing a model or running trials, you must first identify what event you're simulating and what the possible outcomes are.

☑ Practice Test 6 — Answer Key

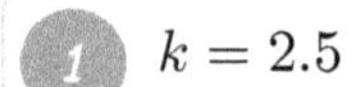 $k = 2.5$  C    C    D    C    C    B    C    C

 A    C   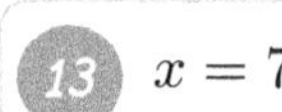 A    $x = 7$  C    B    B    No   18 No

 C    4 m    A    1330 cm^2  D    C    B    B    D

 B    C   30 B

💡 Time to Learn! 💡

Review the explanations below, **especially for the questions you missed**.

Understanding why each answer is correct builds stronger problem-solving skills.

Tip: Circle any questions you got wrong, then read their explanation carefully.

📖 Practice Test 6 — Detailed Explanations

1 $k = \frac{y}{x} = \frac{15}{6} = 2.5$.

2 $k = \frac{10}{4} = 2.5$. When $x = 7$: $y = 2.5 \times 7 = 17.5$.

3 $\frac{3}{87} = \frac{x}{435}$. Cross-multiply: $87x = 1{,}305$, so $x = 15$ gallons.

Find more at
ViewMath.com/MN-Grade7

4 Divide the part by the percent: $45 \div 0.75 = 60$.

5 Planning fee: $4{,}000 \times 0.10 = \$400$. Coordination fee: $4{,}000 \times 0.03 = \$120$. Total: $400 + 120 = \$520$.

6 $\frac{|30-28|}{28} \times 100 = \frac{2}{28} \times 100 \approx 7.1\%$.

7 The interest rate is the percent the lender charges you for borrowing money.

8 On a number line -1 is to the right of -10, so $-1 > -10$.

9 "Divided by 4" gives $\frac{n}{4}$. "Increased by 3" means add 3: $\frac{n}{4} + 3$.

10 Combine variable terms: $-4x + 7x = 3x$. Combine constants: $9 - 3 = 6$. Result: $3x + 6$.

11 Subtract 50: $-4x = -32$. Divide by -4: $x = 8$. Check: $50 - 4(8) = 50 - 32 = 18$ ✓.

12 Area $= 7(x + 3) = 56$. Divide by 7: $x + 3 = 8$. Subtract 3: $x = 5$.

13 Add 7.5: $2.5x = 17.5$. Divide by 2.5: $x = 7$. Check: $2.5(7) - 7.5 = 17.5 - 7.5 = 10$ ✓.

14 Correct steps: subtract 8 ⊠ $-2x > 6$. Divide by -2 and flip: $x < -3$. The student forgot to flip.

15 Subtract 5: $-x > 3$. Multiply by -1 and flip: $x < -3$. Open circle at -3, shade left.

16 All circles with radius 5 cm are congruent — they have the same shape and size. In terms of the figure itself (ignoring position), there is exactly one such circle.

17 $8 + 8 = 16$, which is not greater than 16. The triangle inequality requires the sum to be strictly greater.

Find more at
ViewMath.com/MN-Grade7

18 A rectangular prism has only flat faces and straight edges. All cross-sections are polygons (triangles, rectangles, pentagons, hexagons). A circular cross-section requires a curved surface.

19 Supplementary angles sum to 180°. $110+70=180$ ✓. The others: $45+45=90$, $60+30=90$, $100+100=200$. Only C works.

20 $r^2=12.56\div3.14=4$. So $r=2$ m and $d=4$ m.

21 Room: $14\times10=140$ ft^2. Closet: $3\times4=12$ ft^2. Remaining: $140-12=128$ ft^2.

22 $SA=2(25\times18)+2(25\times5)+2(18\times5)=900+250+180=1330$ cm^2.

23 First: $8\times5\times4=160$ in^3. Second: $10\times4\times4=160$ in^3. They have the same volume.

24 In a random sample, every member of the population has an equal chance of being chosen.

25 Dot plots show every data point, which is useful for small data sets where individual values matter.

26 There are 7 values. The middle value (4th) is 20.

27 55% of $500=0.55\times500=275$.

28 The probabilities are $0.4,0.3,0.2,0.1$ — all different, so this is non-uniform. They sum to 1.0, making it valid.

29 $P(\text{tails})=\frac{1}{2}$ and $P(\text{even})=\frac{3}{6}=\frac{1}{2}$. $P=\frac{1}{2}\times\frac{1}{2}=\frac{1}{4}$. Or: 3 favorable out of 12 total outcomes.

30 A simulation uses random numbers, coins, dice, or spinners to model a real-world event and estimate probabilities.

✅ Practice Test 7 — Answer Key

1. $y = 36$ 2. D   3. 7 hours   4. D   5. $81.60   6. C   7. $60   8. C

9. B   10. A   11. $x = 9$ 12. B   13. A   14. D   15. $x > 4$; open circle at 4, shade right

16. B   17. B   18. B   19. B   20. 39.25 cm^2 21. C   22. C   23. 180 cm^3 24. B

25. Class A   26. B   27. Sports: 40%, 144°; Music: 25%, 90°; Art: 15%, 54°; Reading: 20%, 72°   28. B

29. B   30. C

💡 Time to Learn! 💡

Review the explanations below, **especially for the questions you missed**.

Understanding why each answer is correct builds stronger problem-solving skills.

Tip: Circle any questions you got wrong, then read their explanation carefully.

📖 Practice Test 7 — Detailed Explanations

1. $k = \frac{20}{5} = 4$. So $y = 4 \times 9 = 36$.

2. A proportional relationship always graphs as a straight line through the origin. It cannot be curved.

3. Rate: $42 \div 3 = 14$ mph. Time: $98 \div 14 = 7$ hours.

4. Divide the part by the percent: $9 \div 0.12 = 75$.

5 Fee: $85 \times 0.04 = \$3.40$. Seller receives: $85 - 3.40 = \$81.60$.

6 $\frac{|50-40|}{40} \times 100 = \frac{10}{40} \times 100 = 25\%$.

7 $I = 500 \times 0.24 \times 0.5 = \$60$.

8 The opposite of 0 is 0 because $0 + 0 = 0$. Zero is the only integer that is its own opposite.

9 "Twice a number" is $2n$. "6 more than" means add 6: $2n + 6$.

10 Combine: $\frac{3}{4}n - \frac{1}{4}n = \frac{2}{4}n = \frac{1}{2}n$. Add the constant: $\frac{1}{2}n + 5$.

11 Subtract 100: $-7x = -63$. Divide by -7: $x = 9$. Check: $100 - 7(9) = 100 - 63 = 37$ ✓.

12 The student wrote $3x + 4$ instead of $3x + 12$. Distributing correctly gives $3x + 12 = 30$.

13 Subtract $\frac{1}{2}$: $\frac{3}{4}b = 2\frac{1}{4} = \frac{9}{4}$. Multiply by $\frac{4}{3}$: $b = \frac{9}{4} \times \frac{4}{3} = 3$.

14 "No more than 25" means 25 or fewer. $x \leq 25$.

15 Subtract 4: $3x > 12$. Divide by 3: $x > 4$. Open circle at 4, shade right.

16 All three inequality checks pass ($5 + 5 = 10 > 5$). Three equal sides (SSS) produce exactly one equilateral triangle.

17 SSS with valid lengths produces exactly one unique triangle. Check: $6 + 8 = 14 > 10$, $6 + 10 = 16 > 8$, $8 + 10 = 18 > 6$. All pass.

18 The shape of a cross-section depends on where you cut and at what angle. Different cuts through the same figure can produce different shapes.

19 Complementary angles sum to 90°. $90° - 38° = 52°$.

20 $r = 5$ cm. Full circle: $3.14 \times 25 = 78.5$ cm^2. Semicircle: $78.5 \div 2 = 39.25$ cm^2.

21 Horizontal: $12 \times 2 = 24$ cm^2. Vertical: $2 \times 8 = 16$ cm^2. Total: $24 + 16 = 40$ cm^2.

22 Surface area is the total area covering the outside of a 3D object — the sum of the areas of all its faces.

23 $V = 9 \times 5 \times 4 = 180$ cm^3.

24 Selecting every 10th student from the school roster is systematic and gives all students a chance, making it the least biased.

25 Class A has a smaller IQR (6 vs. 14), meaning the middle 50% of scores are more tightly clustered.

26 A wider range indicates more variability, not better performance. Class A has a higher median and is more consistent (smaller IQR and range).

27 Sports: $\frac{48}{120} = 40\%$, $0.40 \times 360° = 144°$. Music: $\frac{30}{120} = 25\%$, $90°$. Art: $\frac{18}{120} = 15\%$, $54°$. Reading: $\frac{24}{120} = 20\%$, $72°$. Check: $144 + 90 + 54 + 72 = 360°$.

28 A fair number cube has 6 equally likely outcomes, making it a uniform model.

29 $P(heads) = \frac{1}{2}$. Numbers greater than 4: $5, 6$, so $P = \frac{2}{6} = \frac{1}{3}$. $P = \frac{1}{2} \times \frac{1}{3} = \frac{1}{6}$.

30 A spinner with 4 equal sections has $\frac{1}{4} = 25\%$ chance for each section. Assigning one section as "win" gives 25%.

✅ Practice Test 8 — Answer Key

 **1** B    **2** 2.4 miles   **3** 6 cm   **4** D   **5** About $18.60   **6** C   **7** B   **8** C

9 B   **10** Answers vary, e.g., $n + 3n + 7$ or $5n - n + 7$ **11** A   **12** A   **13** $n = \dfrac{16}{5}$ or $3\dfrac{1}{5}$

14 $9s + 35 \leq 200;\ s \leq 18$ **15** $x > 2$; open circle at 2, shade right   **16** 15   **17** AAA   **18** C

19 100° and 80°   **20** B   **21** 67.74 ft^2 **22** B   **23** C   **24** 360

25 Yes; the means are about 2.5 to 3.3 MADs apart   **26** B   **27** 144°   **28** A   **29** $\dfrac{1}{12}$ **30** B

💡 Time to Learn! 💡

Review the explanations below, **especially for the questions you missed**.

Understanding why each answer is correct builds stronger problem-solving skills.

Tip: Circle any questions you got wrong, then read their explanation carefully.

📖 Practice Test 8 — Detailed Explanations

1 $k = \dfrac{150}{2.5} = 60$ miles per hour.

2 Ava: $k = \dfrac{6}{5} = 1.2$ mph; at 6 hr. $1.2 \times 6 = 7.2$ mi. Ben: $k = \dfrac{4}{5} = 0.8$ mph; at 6 hr. $0.8 \times 6 = 4.8$ mi. Difference: $7.2 - 4.8 = 2.4$ miles.

3 $\dfrac{1}{4.5} = \dfrac{x}{27}$. Cross-multiply: $4.5x = 27$, so $x = 6$ cm.

4 $150\% = 1.50$. Multiply: $1.50 \times 40 = 60$. A percent greater than 100% gives a result greater than the whole.

5 Round to $93. Find 10%: $9.30. Double for 20%: $9.30 \times 2 = \$18.60$.

6 $\frac{|35-40|}{40} \times 100 = \frac{5}{40} \times 100 = 12.5\%$.

7 You owe $200 for 1 year. Interest $= 200 \times 0.10 = \$20$.

8 Both $|10| = 10$ and $|-10| = 10$. Two numbers have the same absolute value: the number and its opposite.

9 Substitute: $-4(3) + 10 = -12 + 10 = -2$.

10 Any expression whose like terms combine to give $4n + 7$ is correct. For example: $n + 3n + 7 = 4n + 7$.

11 The pattern is $3x + 5$. Solve $3x + 5 = 41$: subtract 5 to get $3x = 36$, divide by 3: $x = 12$. Check: $3(12) + 5 = 41$ ✓.

12 Divide by 5: $n + 3 = -2$. Subtract 3: $n = -5$. Check: $5(-5 + 3) = 5(-2) = -10$ ✓.

13 Multiply every term by 6: $5n + 2 = 18$. Subtract 2: $5n = 16$. Divide by 5: $n = \frac{16}{5} = 3\frac{1}{5}$.

14 Total cost: $9s + 35 \leq 200$. Subtract 35: $9s \leq 165$. Divide by 9: $s \leq 18.\overline{3}$. Since s must be a whole number, $s \leq 18$.

15 Subtract 10: $-4x < -8$. Divide by -4 and flip: $x > 2$. Open circle at 2, shade right.

16 The third side must be less than $7 + 9 = 16$. The largest whole number less than 16 is 15.

17 Three angles determine the shape but not the size. Infinitely many similar triangles can share the same three angle measures.

Find more at
ViewMath.com/MN-Grade7

18 A cone has a circular base, and all horizontal cuts parallel to the base produce circles of varying sizes.

19 $5x + 4x = 180$. Combine: $9x = 180$. Solve: $x = 20$. Angles: $5(20) = 100°$ and $4(20) = 80°$.

20 The circle's diameter equals the side of the square: $d = 10$, so $r = 5$ cm. $A = \pi r^2 = 3.14 \times 25 = 78.5$ cm^2.

21 Patio: $12 \times 8 = 96$ ft^2. Hot tub: $\pi r^2 = 3.14 \times 9 = 28.26$ ft^2. Remaining: $96 - 28.26 = 67.74$ ft^2.

22 $SA = 2(5 \times 3) + 2(5 \times 2) + 2(3 \times 2) = 30 + 20 + 12 = 62$ cm^2.

23 $V = s^3 = 5^3 = 125$ m^3.

24 $\frac{6}{50} = 12\%$. Then 12% of $3{,}000 = 360$.

25 Difference $= 10$. Using MAD of 4: $\frac{10}{4} = 2.5$. Using MAD of 3: $\frac{10}{3} \approx 3.3$. Both are greater than 2, so the difference is meaningful.

26 $Q_1 = 22$, $Q_3 = 32$. IQR $= 32 - 22 = 10$.

27 $\frac{24}{60} = 0.40 = 40\%$. Angle $= 0.40 \times 360° = 144°$.

28 Predicted: $\frac{5}{10} = 0.5$. Experimental: $\frac{24}{40} = 0.6$. The small difference is expected from natural variation.

29 $P(\text{tails}) = \frac{1}{2}$, $P(1) = \frac{1}{6}$. $P = \frac{1}{2} \times \frac{1}{6} = \frac{1}{12}$.

30 More trials produce more reliable estimates. The experimental probability approaches the theoretical value as trials increase.

Find more at
ViewMath.com/MN-Grade7

☑ Practice Test 9 — Answer Key

| 1 | A | 2 | B | 3 | C | 4 | C | 5 | C | 6 | C | 7 | B | 8 | 500 ft | 9 | A |

| 10 | $12x + 8$ | 11 | B | 12 | A | 13 | C | 14 | C | 15 | A | 16 | B | 17 | C | 18 | C |

| 19 | D | 20 | C | 21 | B | 22 | C | 23 | B | 24 | C | 25 | D | 26 | B | 27 | C | 28 | B |

29 C **30** Roll the die; let 1 and 2 represent the event (or any 2 numbers out of 6)

💡 Time to Learn! 💡

*Review the explanations below, **especially for the questions you missed**.*

Understanding why each answer is correct builds stronger problem-solving skills.

Tip: *Circle any questions you got wrong, then read their explanation carefully.*

📖 Practice Test 9 — Detailed Explanations

1 By convention, $k = \frac{y}{x} = \frac{20}{8} = 2.5$. Student B computed $\frac{x}{y}$, which is the reciprocal — not k.

2 The point $(1, 5)$ means $x = 1$ ticket and $y = \$5$, so one ticket costs $5. This is the unit rate.

3 Rate: $120 \div 8 = 15$ pages/min. Time for 450: $450 \div 15 = 30$ min.

4 Divide the part by the whole: $14 \div 56 = 0.25 = 25\%$.

5 $55 \times 1.20 = \$66.00$.

6. $\frac{|8,000-7,500|}{7,500} \times 100 = \frac{500}{7,500} \times 100 \approx 6.67\%$.

7. Bank A: $2,000 \times 0.015 = \$30$. Bank B: $2,000 \times 0.04 = \$80$. Difference: $80 - 30 = \$50$.

8. Mountain is at 300 ft and Valley is at -200 ft. Difference $= 300 - (-200) = 300 + 200 = 500$ ft.

9. $5n$ means "5 times a number." Subtracting 8 means "8 less than" that product.

10. Perimeter $= 2(4x - 1) + 2(2x + 5) = 8x - 2 + 4x + 10 = 12x + 8$.

11. Substitute $x = 5$: $4(5) + 3 = 20 + 3 = 23$. So $4x + 3 = 23$ is correct.

12. Divide both sides by 2: $x + 6 = 10$. Subtract 6: $x = 4$. Check: $2(4 + 6) = 2(10) = 20$ ✓.

13. Sale price is $\frac{2}{3}p$. After the \$5 coupon: $\frac{2}{3}p - 5 = 15$.

14. Subtract 6: $-3n \leq -6$. Divide by -3 and flip the sign: $n \geq 2$.

15. Add 3: $\frac{x}{2} > 4$. Multiply by 2: $x > 8$. Open circle at 8, shade right.

16. A square with area 36 cm^2 has side 6 cm. All squares with side 6 cm are congruent, so exactly one square satisfies this.

17. $4 + 4 = 8 < 9$. The two shorter sides cannot reach across the longest side.

18. Every cross-section of a sphere is a circle. The size of the circle depends on where you cut, but the shape is always a circle.

19 Let the supplement be x. The angle is $4x$. Sum: $x + 4x = 180$. So $5x = 180$ and $x = 36°$. The angle: $4(36) = 144°$.

20 Full circle area $= \pi r^2 = 3.14 \times 64 = 200.96 \ cm^2$. Quarter circle $= 200.96 \div 4 = 50.24 \ cm^2$.

21 Rectangle: $6 \times 4 = 24 \ cm^2$. Triangle: $\frac{1}{2} \times 6 \times 2 = 6 \ cm^2$. Total: $24 + 6 = 30 \ cm^2$.

22 A cube has 6 faces. $SA = 6s^2 = 6 \times 16 = 96 \ cm^2$.

23 $V = Bh = 24 \times 7 = 168 \ cm^3$.

24 The sample is the smaller group chosen from the population — the 100 students measured.

25 Class A median $= 78$, Class B median $= 65$. Class A's range (60–95) overlaps with Class B's range (50–80), but Class A is higher overall.

26 Group A range $= 20$, Group B range $= 60$. Group B's data is far more spread out.

27 $\frac{3}{24} = 0.125 = 12.5\%$.

28 Predicted: $\frac{1}{4} = 0.25$. Observed: $\frac{25}{80} = 0.3125$. They are close, which supports the model.

29 Only one outcome out of 8 is all tails (TTT). $P = \frac{1}{8}$.

30 Choose 2 of the 6 faces to represent the event. $P = \frac{2}{6} = \frac{1}{3}$.

☑ Practice Test 10 — Answer Key

Find more at
ViewMath.com/MN-Grade7

ViewMath.com

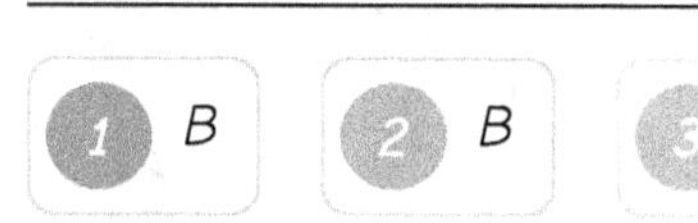 **1** B    **2** B   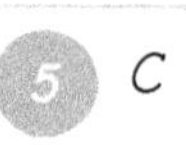 **3** C   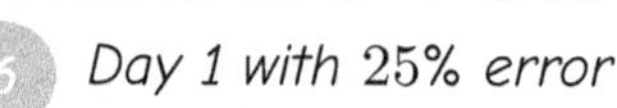 **4** C   **5** C   **6** Day 1 with 25% error   **7** $48

8 0 meters (sea level)   **9** B   **10** B   **11** $x = 13$ **12** B   **13** $6   **14** C   **15** A

16 No   **17** C   **18** Rectangle; 8 cm by 9 cm   **19** B   **20** B   **21** B   **22** C   **23** 6 cm

24 C   **25** B   **26** Median = 300; IQR = 200   **27** B   **28** B   **29** B   **30** A

💡 Time to Learn! 💡

Review the explanations below, **especially for the questions you missed**.

Understanding why each answer is correct builds stronger problem-solving skills.

Tip: Circle any questions you got wrong, then read their explanation carefully.

📖 Practice Test 10 — Detailed Explanations

1 Use any labeled point: $k = \frac{2}{3}$ from $(3, 2)$. Check: $\frac{4}{6} = \frac{2}{3}$ ✓.

2 The point $(1, 9)$ directly gives the unit rate: 9 miles per hour. Check: $\frac{27}{3} = 9$ and $\frac{45}{5} = 9$ ✓.

3 Set up a proportion: $\frac{2}{15} = \frac{7}{d}$. Cross-multiply: $2d = 105$, so $d = 52.5$ miles.

4 $35\% = 0.35$. Multiply: $0.35 \times 480 = 168$ fiction books.

5 Each share: $70 \div 2 = \$35$. Tip: $35 \times 0.15 = \$5.25$. Total each: $35 + 5.25 = \$40.25$.

6 Day 1: $|10 - 8|/8 = 25\%$. Day 2: $|6 - 5|/5 = 20\%$. Day 3: $|8 - 10|/10 = 20\%$. Day 1 has the largest percent error at 25%.

7 $4 \times 12 = \$48$ saved per year.

8 $-200 + 200 = 0$. Rising 200 meters from -200 meters brings the submarine to sea level.

9 Substitute: $2(-3) + 7 = -6 + 7 = 1$.

10 Combine variable terms: $3y - 5y = -2y$. Combine constants: $8 - 2 = 6$. Result: $-2y + 6$.

11 The model shows $3x + 6 = 45$. Subtract 6: $3x = 39$. Divide by 3: $x = 13$.

12 Distribute: $8x - 6 = 26$. Add 6: $8x = 32$. Divide by 8: $x = 4$. Check: $2(4 \cdot 4 - 3) = 2(13) = 26$ ✓.

13 Student $C = \frac{1}{2} \times 12 = \$6$. Check: Student $B = 31.50 - 12 - 6 = \$13.50$. Total: $12 + 13.50 + 6 = 31.50$ ✓.

14 Subtract 9: $-x \leq 6$. Multiply by -1 and flip: $x \geq -6$.

15 Subtract 6: $-2x \leq -10$. Divide by -2 and flip: $x \geq 5$. Closed circle at 5, shade right.

16 $6 + 3 = 9 < 10$. The two shorter sides cannot span the longest side. The triangle inequality is not satisfied.

17 Although $90 + 100 + (-10) = 180$, every angle in a triangle must be greater than $0°$. A negative angle is impossible.

18 A vertical cut through the center of a cylinder produces a rectangle. Width = diameter = $2 \times 4 = 8$ cm. Height = 9 cm.

19 Let the smaller be x. Then $x + (x + 25) = 180$. Combine: $2x + 25 = 180$. Solve: $2x = 155$, $x = 77.5°$. The other: $77.5 + 25 = 102.5°$.

Find more at
ViewMath.com/MN-Grade7

ViewMath.com

20 $A = \pi r^2 = \frac{22}{7} \times 49 = 154\ in^2$.

21   Rectangle: $10 \times 4 = 40\ cm^2$. Triangle: $\frac{1}{2} \times 10 \times 3 = 15\ cm^2$. Total: $40 + 15 = 55\ cm^2$.

22 $SA = 2(30 \times 20) + 2(30 \times 6) + 2(20 \times 6) = 1200 + 360 + 240 = 1{,}800\ cm^2$.

23 $360 = 12 \times w \times 5 = 60w$. $w = 360 \div 60 = 6\ cm$.

24   Average the two samples: $\frac{18+22}{2} = 20$ out of $50 = 40\%$. Then 40% of $1{,}000 = 400$.

25   The median is less affected by skew than the mean, so it is the best measure to compare a skewed and a symmetric distribution.

26   Median (3rd value) $= 300$. $Q_1 = 200$, $Q_3 = 400$. $IQR = 400 - 200 = 200$.

27   The entire circle represents the whole — 100% of the data or $360°$.

28   Red covers half the spinner ($180°$), while Blue and Green each cover a quarter ($90°$ each). This is a non-uniform model.

29   Pairs with product 12: $(2,6), (6,2), (3,4), (4,3)$ — that is 4 out of 36. $P = \frac{4}{36} = \frac{1}{9}$.

30 $P(\text{sum} = 7) = \frac{6}{36} = \frac{1}{6}$. Expected: $36 \times \frac{1}{6} = 6$. The simulation gave 8, slightly above expected.

Well done checking your answers!

Keep practicing to strengthen your skills.

Find more at
ViewMath.com/MN-Grade7

Author's Final Note

I hope you enjoyed this book as much as I enjoyed writing it. Whether you are a student working through the material, a parent supporting your child's learning, or a teacher guiding your class, I have tried to make this book as clear and engaging as possible. I hope I have succeeded. If you have any suggestions for improvement, please let me know. I would love to hear from you.

The accuracy of calculations is very important to me. We have done our best, but I also expect that I have made some minor errors. Constant improvement is the name of the game. If you find any errors, please let me know. I will fix them in the next edition.

For students: Your learning journey does not end here. I have written a series of books to help you learn math. Make sure you browse through them. I especially recommend workbooks and practice tests to help you prepare for your exams.

For parents: Thank you for investing in your child's education. I encourage you to explore the companion resources available online to help support your child outside the classroom.

For teachers: Thank you for the invaluable work you do every day. I hope this book serves as a useful resource in your classroom. Feel free to reach out if you have suggestions or would like to discuss how best to use this book with your students.

I also enjoy reading your reviews. If you have a moment, please leave a review on where you found this book. It will help others find this book. If you have any questions or comments, please feel free to contact me at Nazari@ViewMath.com.

And one last thing: Remember to use online resources for additional help. I recommend using the resources on https://ViewMath.com You can find video lessons, practice problems, and more. You can also use the online companion for this book to track your progress and access additional resources.

Wishing all students the best in their studies, parents every success in supporting their children, and teachers continued inspiration in their classrooms!

Dr. A. Nazari

 # Great Job! Keep Learning with ViewMath!

Keep up the great work! Visit **viewmath.com/MN-Grade7** for free lessons, quizzes, and more.

Study Guide

Workbook

Step-by-Step

3 Practice Tests

5 Practice Tests

7 Practice Tests

Find more at
ViewMath.com/MN-Grade7

ViewMath.com